Genomes and the Gene Pool

Genomes and the Gene Pool

The author, who did much to develop resources that led to the discovery of numerous disease-causing genes and was the first to suggest an approach that was used to sequence the human genome, has written a remarkable book that will be interesting and useful to both lay persons and healthcare providers. The book provides an insightful exploration of human and medical genetics, expertly combining foundational concepts with practical applications such as clinical and prenatal DNA testing. The book includes some controversial topics and opinions which not all readers will agree with, but which will give all pause for thought. This work is a comprehensive resource for anyone interested in understanding the intricate relationships between genetics, health and society.

—**Val Sheffield, MD, PhD**
Professor of Pediatrics and Ophthalmology and Visual Sciences
Roy J. Carver Chair in Molecular Genetics, University of Iowa
March 27, 2026

After providing a thorough description of human genetics, Weber then discusses the evolution of human intelligence, the historic factors and events that might have led to the emergence of our "special intelligence." In a thoughtful, humane and readable way he raises questions and expresses views around the future evolution of human intelligence. Here's the stage on which the rise of artificial intelligence will play out.

—**John Todd, PhD**
Professor of Precision Medicine, University of Oxford
February 25, 2026

Having been at the forefront of the genomics revolution for over five decades, in *Genomes and Gene Pool* Jim Weber offers a deeply personal overview, reflection, and analysis of the major advances that have shaped modern genetics. Moving beyond the scope of a conventional textbook, the book blends insight with provocation, challenging assumptions and stimulating debate, while remaining firmly grounded in empirical evidence.

—**Anne Goriely, PhD**
Professor of Human Genetics, University of Oxford
April 2, 2026

Genomes and the Gene Pool

How Genetics is Revolutionizing Healthcare and Reproduction

JAMES L. WEBER

TEN GENERATIONS PRESS

Published by Ten Generations Press, Marshfield, Wisconsin

www.tengenerationspress.org

Genomes and the Gene Pool:
How Genetics is Revolutionizing Healthcare and Reproduction

Publisher's Cataloging-in-Publication
(Provided by Cassidy Cataloguing Services, Inc.)

Names: Weber, James L., author.
Title: Genomes and the gene pool : how genetics is revolutionizing healthcare and reproduction / James L. Weber.
Description: Marshfield, Wisconsin : Ten Generations Press, [2026] | Includes bibliographical references and index.
Identifiers: LCCN: 2026914046 | ISBN: 9798995825326 (hardcover) | 9798995825319 (paperback) | 9798995825302 (ebook)
Subjects: LCSH: Medical genetics. | Genomics. | Medical care. | Family planning. | Human genetics. | Human evolution. | BISAC: SCIENCE / Life Sciences / Genetics & Genomics. | MEDICAL / Genetics. | SCIENCE / Life Sciences / Evolution.
Classification: LCC: RB155 .W43 2026 | DDC: 616.042--dc23

TABLE OF CONTENTS

LIST OF FIGURES

LIST OF TABLES

LIST OF CASES

ABBREVIATIONS

A	Adenine
ADR	Adverse drug reaction
AI	Artificial insemination
ARTs	Assisted reproductive technologies
ASDs	Autism spectrum disorders
BCE	Before Common Era
C	Cytosine
CE	Common Era
CF	Cystic fibrosis
CVS	Chorionic villus sampling
DNA	Deoxyribonucleic acid
ELSI	Ethical, legal, and social implications (of genetics)
FBI	Federal Bureau of Investigation (US)
FDA	Food and Drug Administration (US)
G	Guanine
HLA	Human leukocyte antigen
ICSI	Intracytoplasmic sperm injection
ID	Intellectual disability
IVF	In vitro fertilization
MH	Malignant hyperthermia
mRNA	Messenger RNA
NHS	National Health Service (UK)
NICU	Neonatal intensive care unit
NIH	National Institutes of Health (US)

NIPS	Non-invasive prenatal screening
PGT	Preimplantation genetic testing
PKU	Phenylketonuria
RNA	Ribonucleic acid
T	Thymine
U	Uracil
UK	United Kingdom
US	United States
WWII	World War II

PREFACE

A little about my background may be helpful to the reader. I received a PhD in biochemistry from the University of California–Berkeley in 1980. For nearly all of the subsequent 45 years, I worked in the field of human genetics. In roughly the first half of this period, I served as head of a large research lab. I competed for research grants, carried out research studies, attended research conferences, and authored papers for scientific journals. In the second half of my career, I served as the founder and head of a clinical DNA testing laboratory. Therefore, regardless of my other inadequacies, I have over my career accumulated a great deal of knowledge and experience in human genetics.

I included a good number of references in this book for two purposes: to provide support for many of the statements made in the text, and as a source of additional information for those who would like to learn more. Consistent with the basic level of genetics material in this book, I avoided comprehensive lists of references and, where possible, excluded particularly technical references. I also generally referenced more recent articles, which include references to earlier work. The references are numbered separately for each chapter.

I also tried to reference articles that are freely available through the internet. Such "open source" references are marked with [Open] in the reference lists. The easiest way to view the open source articles is through PubMed, a service of the US government's National Library of Medicine. To access references with PubMed, simply go to the PubMed website and enter into the search box the ID numbers found in the reference lists. To access articles that are not open source, readers will probably need to visit a college or university library.

Nine clinical cases involving individual patients and families are presented in this book. In the interest of privacy, all of these cases are fictional. They are based, however, on many similar cases that I have encountered over my career.

I minimized the use of abbreviations in this book, but was unable to avoid them entirely. Readers may therefore be aided by the list of abbreviations found near the beginning of the book, just before this preface.

In a few places in the book, I included costs of DNA testing or healthcare procedures and treatments. All of these costs are in 2025 US dollars.

Artificial intelligence was not used to write any part of this book.

All income from sale of this book will be donated to charity.

Like all scientists, I stand upon the shoulders of thousands of other scientists, and have benefited enormously from the time and generosity of many of my colleagues. I am particularly grateful to my early mentors, David Cole, Jack Gorski, and Wayne Hockmeyer. I also thank Ginny Brooks and Ann MacLaughlin-Berres for expert assistance with the illustrations.

PART I:

GENETICS AND HEALTHCARE

PART I:

GENETICS AND HEALTHCARE

Chapter 1

INTRODUCTION TO PART I

The field of human genetics has advanced rapidly over the last several decades. We have gained unprecedented ability to learn about our own genomes and even to alter the course of human evolution. The decisions we make—or do not make—over the next few generations regarding management of the gene pool will profoundly affect the future of our species.

This book has two parts. Part I provides a basic introduction to human genetics, with emphasis on mutation, heritability of traits, and human evolutionary history. It concludes with two chapters on clinical DNA testing. Clinical DNA testing is already leading to significant improvements in healthcare, particularly through genome sequencing.

Part II deals with the gene pool. Because many readers may find Part II more interesting than Part I, you may be tempted to jump directly to Part II. But I encourage you to soldier through the first part of the book, especially Chapter 2 (things get more interesting beginning in Chapter 3). The basic information in Part I lays an important foundation without which Part II cannot be fully understood and appreciated.

This book was written primarily for educated individuals without a background in genetics or even in science. Although the book was not tailored for healthcare providers, many providers may find it helpful, especially those who have not received advanced training in genetics.

Even professional geneticists may find portions of the book useful, particularly the later chapters of Parts I and II. To assist readers, I included numerous figures and tables, as well as brief, bullet-point summaries at the end of each chapter. The summaries are not substitutes for the text, but provide a concise review of content. Given the intended audience, I focused on basic concepts and generally avoided complexities and rare exceptions to general rules. I refer those who would like to delve deeper into this subject to some excellent recent textbooks on human genetics (1–4).

All books about rapidly advancing scientific fields like human genetics become, at least to a degree, outdated relatively quickly. Some may think it is folly to even attempt such a book. But I counter that if the only science books ever written were those that would never become outdated, then no science books would ever be written. I'm confident that the basic genetic facts presented in this book, although they will gradually become better understood and elaborated, will stand the test of time. Given the current state of healthcare and reproduction, I also think the time is right for a book about genomes and the gene pool.

Like all scientific disciplines, human genetics has many words or phrases that are frequently used by geneticists, but rarely used by people outside of the field. Learning this jargon is a big part of understanding. Where possible, I avoided jargon. I used, for example, "identical twins" rather than "monozygotic twins," the term usually used by geneticists. However, in many places, I found it impossible to avoid jargon. Therefore, when a word is unfamiliar, you are encouraged to refer to the glossary located after the final chapter. My definitions are relevant and appropriate for this book, but you may also want to refer to other genetic glossaries (they may be found by searching the internet for "glossaries of genetic terms").

This book is not a textbook, but like textbooks, I tried to present as fact only those scientific concepts that are thoroughly supported by evidence and upon which experts concur. I mostly avoided newer, unproven ideas, but where necessary, tried to clearly indicate their unestablished nature. Hopefully, very little that I have presented as fact will turn out to be incorrect.

This book also contains some personal opinions and beliefs, especially, but not exclusively, in Chapters 7–8 and 15–18. Not all questions that are important can be answered by science. No experiments or observations can tell me whether abortion is right or wrong, or which candidate I should vote for in the next election. In my experience, many science writers don't do a good job distinguishing fact from opinion. In this book, I made a strong effort to clearly distinguish my personal opinions from scientific fact using such language as "I think" or "my opinion is."

To fully appreciate this book, it's important to have a basic understanding of the scientific method and scientific truth. Scientists do not, and indeed cannot, change how nature works, but rather seek to understand nature and then communicate this understanding to others. Science advances through the proposal of new ideas, often called hypotheses or theories. A new idea is always received with skepticism by other scientists, who carefully scrutinize the evidence in favor or against. Since scientists are at least as fallible as any other group of professionals, many new ideas turn out to be wrong. Only gradually, usually over a period of at least several years and often decades—if no one has found a flaw in the idea, if evidence in support of it has grown substantially, and if it is independently confirmed by others—will the idea be accepted as scientific truth.

When a new idea moves from hypothesis to established fact, it works its way from scientific conferences and scientific journals to textbooks. It also moves from a proposal by a single scientist or small group of scientists to consensus among all or nearly all scientists within the field. Almost all the information found in good science textbooks stands the test of time. Textbooks on human genetics written over 50 years ago (a long time for a rapidly moving field such as human genetics) are still useful and almost entirely correct (see, for example, reference 5). However, the concepts found in textbooks are constantly being extended and improved. So, the newest textbooks generally contain the most up-to-date information.

Having worked for many years in both nonprofit research and private industry, I have learned that nonprofit scientific research is a very competitive activity, generally more competitive even than private industry. Nearly every budding young research scientist dreams of becoming a professor at a major research university and perhaps winning a Nobel Prize, but few attain these heights. The most successful research scientists are those who generate during their careers the greatest numbers of correct new ideas and the most valuable new ideas. Incorrect ideas are shredded by competitors.

Truth is often a squishy concept, but one aspect of truth that is rock-solid is the way nature works. How do we know that established scientific concepts are true? First, competitors and critics have tried hard to find flaws, but have been unsuccessful (6). And second, and more important, science is eminently useful. Science has achieved amazing success in advancing human civilization, particularly over about the last 200 years. No other human endeavor has come even remotely close to science in advancing humanity during this period (7).

Human genetics is sometimes an unpleasant subject. It's frequently linked to people with serious health problems or disabilities. In some ways, genetics is similar to phenomena such as hurricanes and earthquakes. We don't like such occurrences, but we can't stop them. I believe that we are far better off if we try to understand and prepare for such occurrences than if we ignore them. This book isn't gloomy, but I did attempt to present the unvarnished realities of human individual and population genetics. Ignoring such realities is dangerous.

But human genetics also often leads to great victories. Patients are brought back from the brink of death. New cures for disease are found. Many disorders and disabilities are prevented. Understanding and appropriately utilizing genetics will lead us to a brighter future.

Chapter 2

HUMAN GENETICS BASICS

Case 1: Back from the Brink

Maria and Dan were overjoyed at the birth of their first child. Initially the new baby, a boy named Alesandro, seemed perfectly healthy. But before Alesandro left the hospital, he developed seizures. Despite intense efforts, the doctors were unable to control the seizures. They didn't have a diagnosis. The outlook for Alesandro was grim.

As a last resort, Alesandro's doctors decided to try DNA testing. A small blood sample was drawn and shipped to a clinical laboratory. In the lab, Alesandro's DNA was extracted from the blood, and all of his genes were sequenced. Analysis of the sequences revealed that Alesandro was heterozygous for two deleterious variants in the *ALDH7A1* gene. As soon as Alesandro's doctors received this result, they placed him on relatively large doses of vitamin B6. Alesandro improved rapidly and was able to go home with his parents in a few days.

Most likely, you don't yet understand all the concepts and jargon presented in Case 1, but by the time you finish Chapters 2–4, you should be able to understand this case.

The Role of DNA

People vary in many ways: some are tall, some short; some have high blood cholesterol, some low; some have excellent vision, some poor. Human genetics is the science of the causes and the inheritance of human variation. You've heard that genetics is connected with something called DNA. Let's see how this works.

First, let's break our bodies down into smaller and smaller parts (Figure 2.1). You know that inside our bodies we have organs and tissues such as brain, kidneys, and blood. You also probably know that each of these organs and tissues is in turn composed of billions of tiny, membrane-enclosed units called cells. But now let's go beyond the cellular level, even beyond the subcellular level, down to the molecular level. We have many different types of molecules in our bodies. Some are relatively small molecules such as sugars and fats. But what distinguishes living material from, say, a soft drink are not the small molecules, but rather the large or, as biochemists call them, the macromolecules. DNA, RNA, and proteins are all macromolecules.

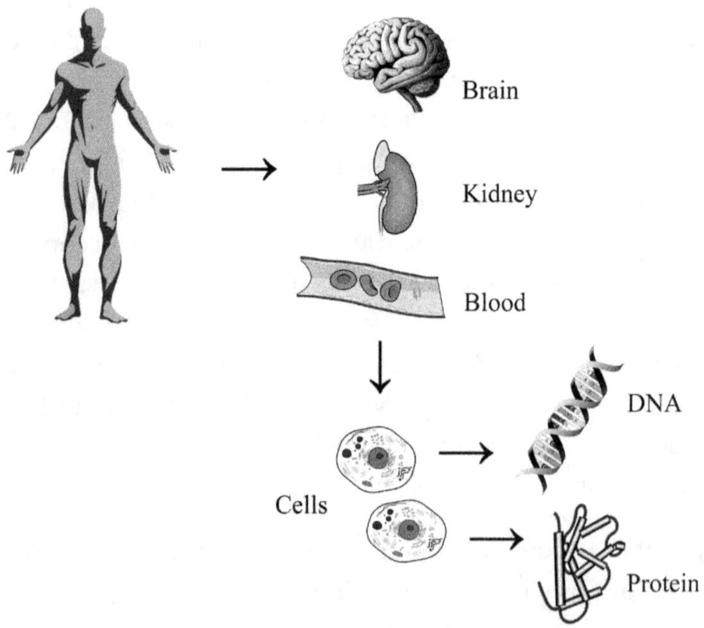

Figure 2.1 Breaking our bodies down into smaller and smaller parts

Of the macromolecules in our bodies, the most functionally diverse are proteins. Proteins are the structural components that comprise much of muscle, skin, and hair. Proteins are also the enzymes that catalyze the myriad chemical reactions that take place in our bodies. And many of the signaling molecules that convey information both between and within cells are proteins. We have about 20,000 basic types of proteins in our bodies (1), and because proteins are tiny, many, many copies of each type. For example, the protein hemoglobin, which ferries oxygen in our blood, is present in our bodies in about a billion trillion (1,000,000,000,000,000,000,000) copies.

Some proteins, often called housekeeping proteins, are manufactured in nearly all cell types, but the majority of our 20,000 proteins are manufactured in only some organs and tissues (2). Our hearts, for instance, contain special muscle proteins that are present only in the heart, and our most complex organ, the brain, contains many proteins that are specific to brain (3).

Most protein molecules have relatively short lifetimes. They last only hours or days until they cease functioning and are recycled (4). Consequently, new proteins need to be continuously manufactured. This is where DNA comes in. While also a macromolecule, DNA does not comprise any structures or catalyze any reactions. Yet DNA contains all the *information* necessary to manufacture new proteins. The basic flow of information in our cells is from DNA to an RNA intermediate to protein (Figure 2.2) (5). This is one of the most fundamental principles in all of biology.

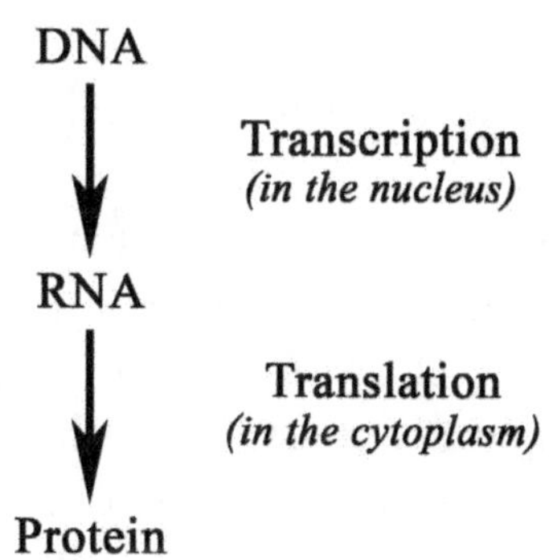

Figure 2.2 Flow of information within cells

Now let's consider our bodies from a different viewpoint. As amazing as it sounds, all human life begins as a single cell, the fertilized egg cell. A human egg cell is about the size of a small period at the end of a sentence. As cells go, it's quite large, but compared to even a baby, it's obviously tiny. A human egg cell as observed through a microscope is shown in Figure 2.3.

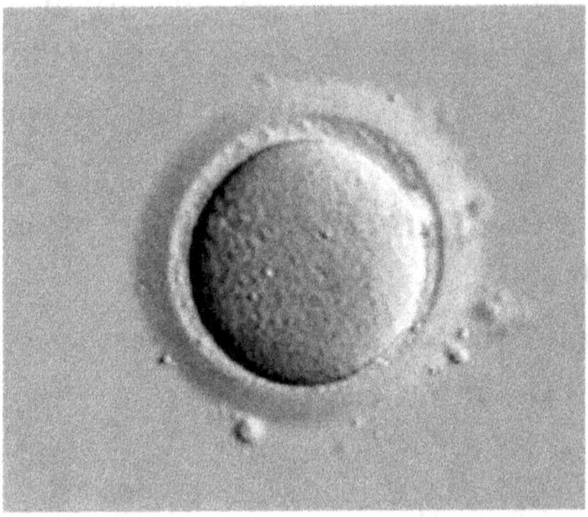

Figure 2.3 Human egg cell viewed through a microscope

Centuries ago, many scholars believed that inside the fertilized egg cell was a tiny person, who grew as the baby developed. We now know that there are no tiny people or brains or bones or blood within the egg cell. However, present within this cell are all the DNA instructions necessary to generate all the organs and tissues in our bodies. In this way, DNA is like a blueprint for a building. The blueprint is not the building itself, but contains all the information necessary to construct the building. The DNA blueprint contains all the information necessary to build a person.

To form a person, the first cell—the fertilized egg cell—divides to produce two cells. Those two cells then divide to form four, then eight, and so on until the trillions of cells in our adult bodies are formed (Figure 2.4) (6). Because of the exponential growth in cell numbers, it only takes about 40 rounds of cell division to go from one to a trillion cells.

Every time a cell divides, the DNA molecules within that cell are copied with exceptional fidelity, and a full set of the DNA instructions is distributed to each daughter cell. As a result, each cell within our bodies contains the full DNA blueprint.

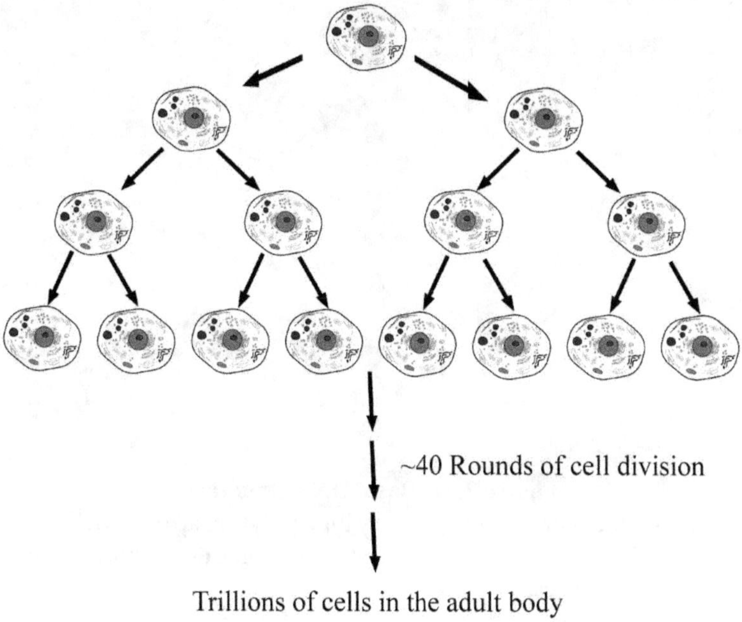

Figure 2.4 Early cell division in prenatal development

DNA Structure and How Protein Is Manufactured

To understand how the information in the DNA molecules is used to manufacture protein, we need to learn a little about the structure of DNA. DNA is a linear, unbranched, chain-like molecule (a type of polymer) composed of repeating units called nucleotides (Figure 2.5). Each nucleotide consists of three small molecules bound together: a phosphate, a type of sugar called deoxyribose, and one of four types of chemical units called bases. The information in DNA is stored as the *sequence* of nucleotides along the polymer.

The four bases are abbreviated by the first letter in their chemical names: A for adenine, C for cytosine, G for guanine, and T for thymine. Before the structure of DNA was determined, chemists learned that in the DNA of any species, including humans, the amount of the A base was always equal to the amount of the T base, and the amount of G base was always equal to the amount of C base (7). This was an important clue in working out the DNA structure.

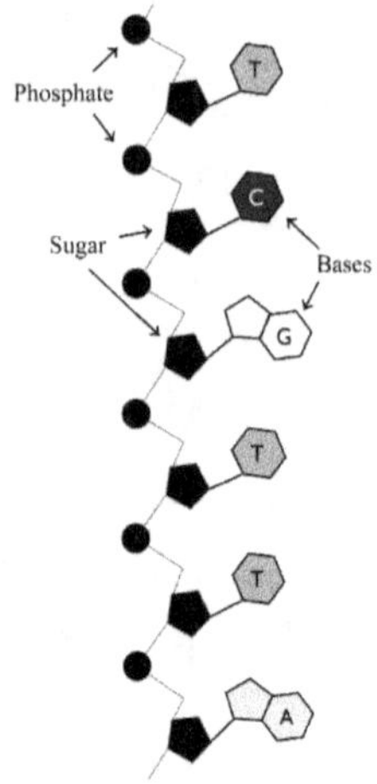

Figure 2.5 Basic DNA structure
Only six nucleotides out of a very long polymer are shown. Each nucleotide consists of a phosphate, the sugar deoxyribose, and one of four bases.

In cells, DNA is not a single-stranded polymer, but rather consists of two different polymer strands bound together. In this double-stranded structure, the bases pair in a special way: A always pairs with T, and G always pairs with C (Figure 2.6). Because of these base-pairing rules, if the order, or, as geneticists say, the sequence, of nucleotides along one of the DNA strands is known, then it's trivially easy to determine the sequence of nucleotides along the other strand. In Figure 2.6, for example, since the sequence of one strand starting from the top is GAAG, the sequence of the other strand must be CTTC. The two strands in DNA are twisted around each other like a spiral staircase. This spiral structure is called a helix, and because there are two strands bound together, the DNA structure is called a double helix.

James Watson and Francis Crick (Figure 2.7) reported the double-helical structure of DNA in the journal *Nature* in 1953 (8). They won a Nobel Prize for this work. Near the end of their article, Watson and Crick placed this famous statement:

> It has not escaped our notice that the specific pairing we have postulated immediately suggests a possible copying mechanism for the genetic material.

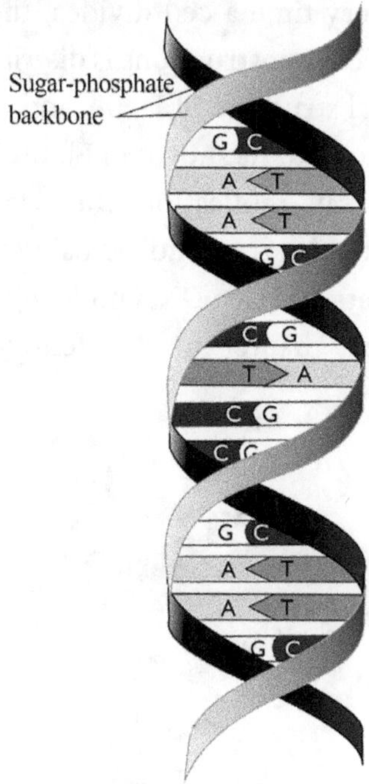

Figure 2.6 Double-helical structure of DNA

Figure 2.7 Watson (left) and Crick with their double-helical model of DNA
Photo credit: A. Barrington Brown / Science Photo

Remember that every time a cell divides, the DNA molecules are copied and a full set of DNA instructions is distributed to each daughter cell. The double-helical structure of DNA, and the base-pairing rules, indicate how DNA is copied. As shown in Figure 2.8, during DNA copying or, as geneticists say, replication, the two original strands split apart, and two new strands are produced using the base-pairing rules. The outcome of replication is two DNA molecules that are identical (or as explained in the next chapter, nearly identical) in sequence to the original DNA molecule.

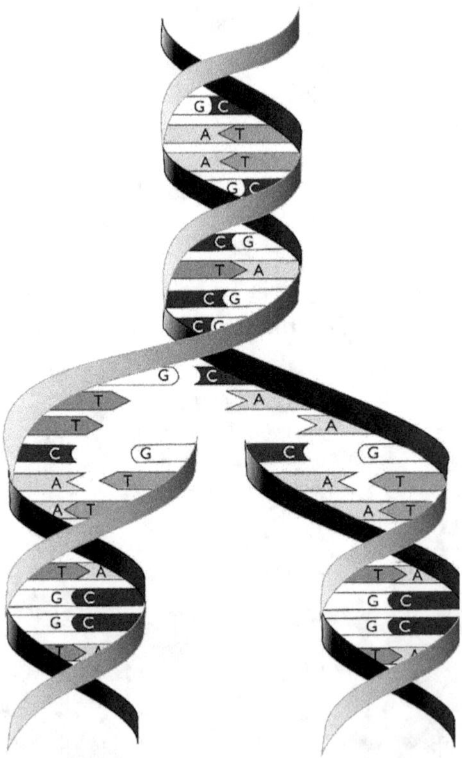

Figure 2.8 Replication of DNA

But how are proteins made? Like DNA, proteins are also unbranched, linear polymers. But unlike DNA, protein polymers have 20, rather than 4, different types of repeating units called amino acids. The names of the 20 different amino acids along with their three-letter abbreviations are shown in Table 2.1.

Table 2.1
Amino acids within proteins

Alanine	Ala	Leucine	Leu
Arginine	Arg	Lysine	Lys
Asparagine	Asn	Methionine	Met
Aspartic acid	Asp	Phenylalanine	Phe
Cysteine	Cys	Proline	Pro
Glutamic acid	Glu	Serine	Ser
Glutamine	Gln	Threonine	Thr
Glycine	Gly	Tryptophan	Trp
Histidine	His	Tyrosine	Tyr
Isoleucine	Ile	Valine	Val

Each of the approximately 20,000 proteins in our bodies has a unique sequence of amino acids. Proteins range greatly in length, from some small proteins with just tens of amino acids to the largest protein called titin with over 35,000 amino acids. The sequence of amino acids in a protein is coded in a simple way by the sequence of nucleotides in DNA.

Since there are 20 different amino acids but only four nucleotides, one nucleotide can't code for one amino acid. This would provide only 4 codes. And two adjacent nucleotides are also insufficient, since this would provide only 16 codes. Rather, as shown in Figure 2.9, three adjacent nucleotides code for each amino acid along the protein polymer. Figure 2.9 is the genetic code, which is universal among all living things on Earth. Each amino acid has at least 1 three-nucleotide code, and most amino acids have 2 to 6 redundant codes. As an example, the amino acid proline (abbreviated Pro) is coded by CCT, CCC, CCA, and CCG. There are also 3 codes for "Stop" (TAA, TAG, and TGA). When the protein-manufacturing machinery hits a Stop code, it completes production of the protein.

TTT	*Phe*	TCT	*Ser*	TAT	*Tyr*	TGT	*Cys*
TTC	*Phe*	TCC	*Ser*	TAC	*Tyr*	TGC	*Cys*
TTA	*Leu*	TCA	*Ser*	TAA	*Stop*	TGA	*Stop*
TTG	*Leu*	TCG	*Ser*	TAG	*Stop*	TGG	*Trp*
CTT	*Leu*	CCT	*Pro*	CAT	*His*	CGT	*Arg*
CTC	*Leu*	CCC	*Pro*	CAC	*His*	CGC	*Arg*
CTA	*Leu*	CCA	*Pro*	CAA	*Gln*	CGA	*Arg*
CTG	*Leu*	CCG	*Pro*	CAG	*Gln*	CGG	*Arg*
ATT	*Ile*	ACT	*Thr*	AAT	*Asn*	AGT	*Ser*
ATC	*Ile*	ACC	*Thr*	AAC	*Asn*	AGC	*Ser*
ATA	*Ile*	ACA	*Thr*	AAA	*Lys*	AGA	*Arg*
ATG	*Met*	ACG	*Thr*	AAG	*Lys*	AGG	*Arg*
GTT	*Val*	GCT	*Ala*	GAT	*Asp*	GGT	*Gly*
GTC	*Val*	GCC	*Ala*	GAC	*Asp*	GGC	*Gly*
GTA	*Val*	GCA	*Ala*	GAA	*Glu*	GGA	*Gly*
GTG	*Val*	GCG	*Ala*	GAG	*Glu*	GGG	*Gly*

Figure 2.9 The genetic code

By knowing the sequence of nucleotides in the DNA and the genetic code, it's easy to decipher the amino acid sequence of the coded protein. Shown in Figure 2.10 is the beginning of the DNA, RNA, and amino acid sequences of the beta subunit of the protein hemoglobin. RNA, like DNA, is a linear, unbranched polymer of nucleotides. The structure of RNA is very similar to DNA, except that RNA is single stranded, contains the sugar ribose instead of deoxyribose, and contains the base uracil (U) instead of thymine (T). To produce an RNA molecule, the two strands of the DNA template split apart, much as in DNA replication, and the RNA molecule is generated using essentially the same base-pairing rules as in the DNA double helix (U pairs with A, just like T pairs with A). The sequence information in the DNA is thereby retained in the RNA.

The segment of a DNA molecule that codes for a single protein is known as a gene. Our genomes have about 20,000 genes, with one gene coding for each protein. The gene shown in Figure 2.10 starts with an ATG code (AUG in the RNA). As indicated in the first column of Figure 2.9, ATG codes for the amino acid methionine, or Met for short. The next code, GTG in the DNA and GUG in the RNA (also found in

DNA:	AGCAACCTCA	AACAGACACC	ATG	GTG	CAC	CTG	ACT	CCT	GAG	GAG	AAG	TCT
RNA:	AGCAACCUCA	AACAGACACC	AUG	GUG	CAC	CUG	ACU	CCU	GAG	GAG	AAG	UCU
Protein:			Met	Val	His	Leu	Thr	Pro	Glu	Glu	Lys	Ser

Figure 2.10 Beginning of the sequence of the beta subunit of hemoglobin
Shown are 20 nucleotides prior to the beginning of the gene, followed by 10 three-nucleotide codes for the first 10 amino acids in the protein.

the first column of Figure 2.9), codes for the amino acid valine, or Val for short. The process continues through all portions of the gene until a Stop code is reached.

DNA is located in a part of the cell called the nucleus, often situated near the cell center. RNA molecules are produced in the nucleus. After some important processing of the initial RNA molecule in the nucleus (see below), the RNA molecule, now called a messenger RNA (mRNA), moves from the nucleus to the outer portion of the cell, called the cytoplasm (Figure 2.2). In the cytoplasm, tiny protein-manufacturing machines called ribosomes bind to the mRNA and produce protein.

Genome just means the total complement of DNA molecules within a species or individual. Genome can therefore refer to the human genome (as opposed to the rice genome), but can also refer to Alesandro's (or any other person's) genome.

Chromosomes

The human nuclear genome is composed of 46 DNA molecules of varying lengths, ranging from about 45 million nucleotides to 250 million nucleotides (about 6.2 billion nucleotides in total). The DNA molecules are *much* longer than they are wide. For instance, the DNA molecule in our largest chromosome, chromosome 1, is about 8.5 centimeters long, but is only 2 nanometers wide (a nanometer is one billionth of a meter). In addition, since the typical human cell is only about 30 micrometers in diameter (a micrometer is one millionth of a meter), the length of the chromosome 1 DNA molecule is approximately 3,000 times the cell diameter. Our DNA is therefore highly compacted within our cells. Compaction is achieved by combining the DNA with several different types of protein. The complexes of DNA and protein are called chromosomes. Shown in Figure 2.11 is a cartoon of chromosome structure (left), and an electron microscopic image of a chromosome with most of its proteins removed and the DNA unpacked (right). The "fuzz" around the chromosome scaffold is the DNA within just this one chromosome.

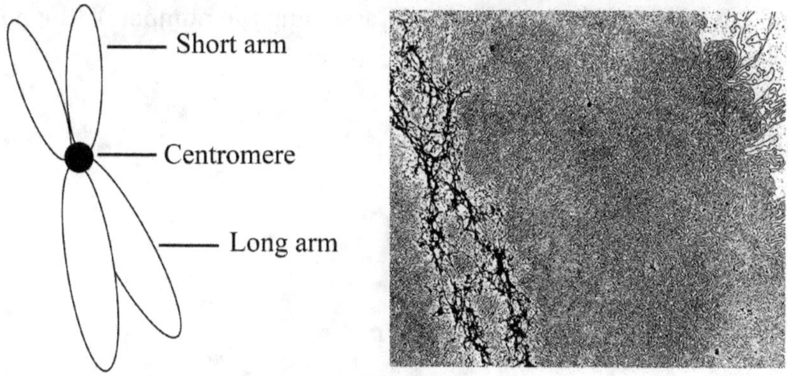

Figure 2.11 Basic chromosome structure (left) and electron microscope image of a partially unraveled chromosome (right)

The 46 human chromosomes are organized into 23 pairs. The nucleotide sequences of DNA molecules within a pair of chromosomes are nearly identical, while the sequences of DNA molecules from different pairs are totally unrelated. For example, in any person, the nucleotide sequences of the DNA molecules from the two copies of chromosome 7 are almost identical, but the sequences of the DNA molecules from one copy of chromosome 7 and one copy of chromosome 19 are completely different.

When cells divide, the chromosomes compact even more than normal. When maximally compacted, they can be conveniently viewed in a microscope. Shown in Figure 2.12 are the compacted chromosomes from one cell, all jumbled. Before viewing through the microscope, chromosomes are typically stained with a dye that produces bands along the chromosome arms. The banding patterns are unique for each chromosome. Using the banding patterns and chromosome length, trained experts, assisted by computer programs, can identify each chromosome. Shown in Figure 2.13 are the chromosomes after each pair has been aligned.

Most of our chromosomes are named using numbers such that chromosome 1 has the longest DNA molecule, and chromosome 22 the shortest (well, actually, chromosome 21 has the shortest, but geneticists goofed when they were assigning the numbers). The numbered chromosomes are called autosomes.

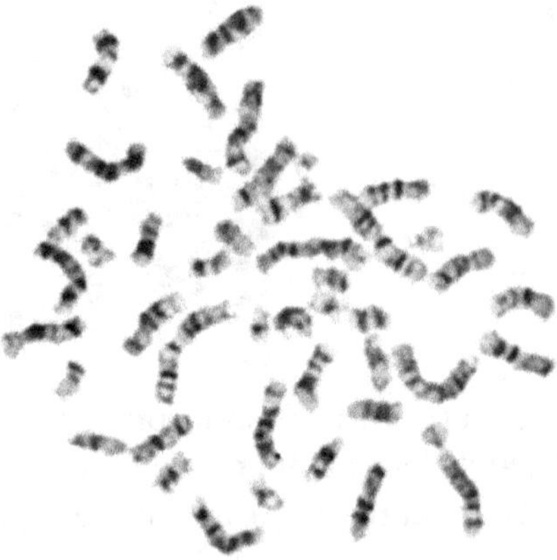

Figure 2.12 Jumbled human chromosomes

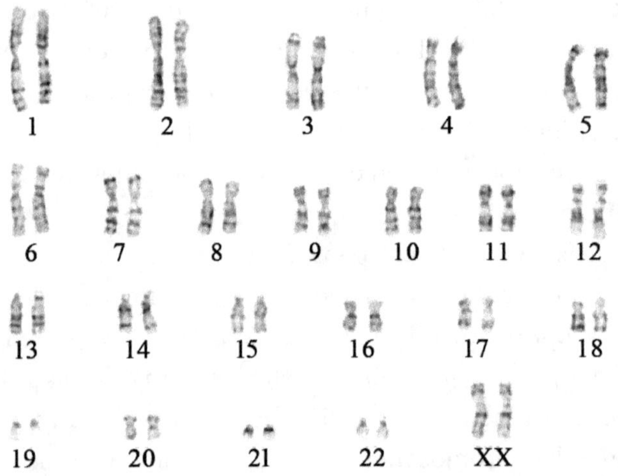

Figure 2.13 Aligned chromosomes from a human female
(from https://biology.stackexchange.com/questions/47944)

In addition to the 22 pairs of autosomes, humans also have two sex chromosomes named X and Y. X is an average-sized chromosome; Y is small. The X and Y chromosomes are called sex chromosomes because they differ between females and males. Females have two X chromosomes and no Y chromosome, while males have one X and one Y. The chromosomes shown in Figure 2.13 are therefore from a female: 22

pairs of numbered chromosomes, two X chromosomes, and no Y. And the aligned chromosomes from a male are shown in Figure 2.14: 22 pairs of numbered chromosomes, one X, and one Y.

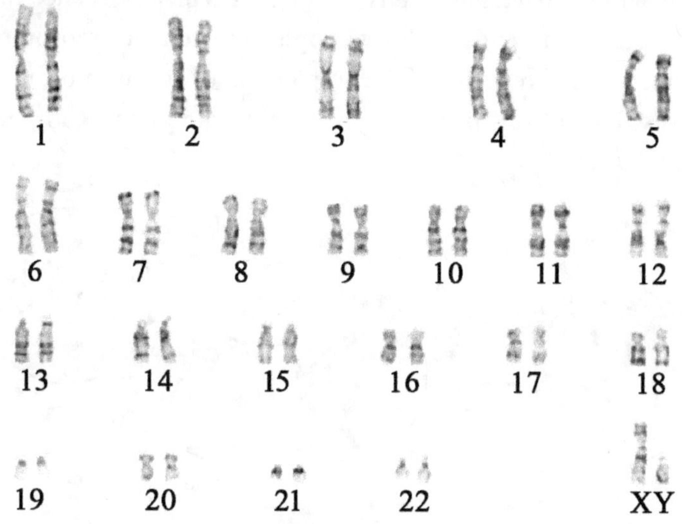

Figure 2.14 Aligned chromosomes from a human male
(from https://biology.stackexchange.com/questions/47944)

Chromosome Distribution During Cell Division

Biologists split the cells in our bodies into two groups: somatic cells and germ cells. Early in pregnancy, at about 16 days post fertilization, the cells within the developing embryo split. The larger group of cells goes on to develop into the somatic cells that are defined as all the cells in our bodies except for the germ cells. Brain, blood, skin, liver, and muscle cells are all somatic cells. The smaller group of cells goes on to develop into the germ cells: the egg cells in the ovaries of females and the sperm cells in the testes in males. DNA within somatic cells is not inherited; only DNA from the germ cells is transmitted to the next generation.

When somatic cells divide, the chromosomes are distributed to the daughter cells as shown in Figure 2.15. For simplicity, only two pairs of chromosomes are shown, but the process is the same for all the other chromosomes. The first step is the replication of each DNA molecule in each chromosome. A cell that is about to divide therefore has double the normal amount of DNA. Then, as the cell divides, the chromosomes split down the middle, and half of each chromosome is pulled into one of the two daughter cells. Through this process, called mitosis, each cell starts and ends with two copies of each chromosome type. Each daughter cell receives a full complement of 46 chromosomes. You can find videos of this process by searching the internet for "mitosis videos."

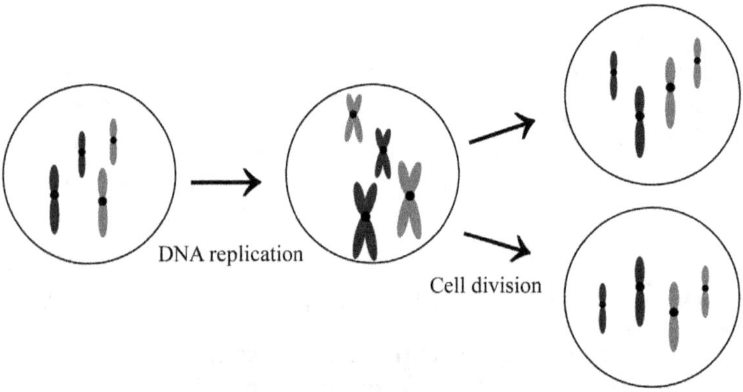

Figure 2.15 Mitosis

Only 2 of the 24 pairs of chromosomes are shown. The chromosomes inherited from the mother of the person in whom mitosis is taking place are shown in light gray, and the chromosomes from the father in dark gray. After DNA replication, the original DNA molecule and the replicate are bound together at chromosome constrictions called centromeres (Figure 2.11). To better illustrate the principles of mitosis, the chromosomes are shown only in fully compacted form. In actual cells, DNA replication takes place before the chromosomes fully compact.

During formation of the egg and sperm cells, and only during the formation of these cells, distribution of chromosomes occurs through a different process called meiosis. Just as in mitosis, the first step is replication of the DNA molecules to give double the amount of DNA, but then meiosis and mitosis diverge. In meiosis, as shown in Figure 2.16, cells divide twice, rather than just once, as in mitosis. In the first meiotic cell division, the two replicated chromosomes of each type align, and one of each type goes to each of the daughter cells.

Next, a second cell division occurs that is much like the cell division in mitosis; one half of each replicated chromosome goes to each daughter cell. When meiosis is complete, each germ cell now has only one of each chromosome type. Egg and sperm cells therefore contain 23 rather than 46 chromosomes and half the normal amount of DNA. You can find videos of this process by searching the internet for "meiosis videos."

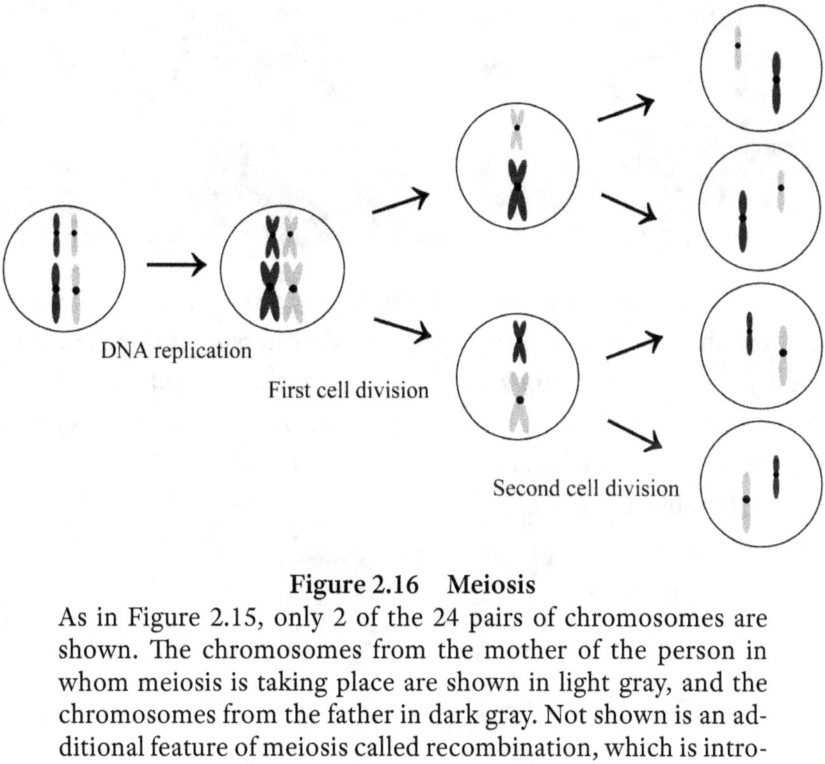

Figure 2.16 Meiosis

As in Figure 2.15, only 2 of the 24 pairs of chromosomes are shown. The chromosomes from the mother of the person in whom meiosis is taking place are shown in light gray, and the chromosomes from the father in dark gray. Not shown is an additional feature of meiosis called recombination, which is introduced in Chapter 4.

When the sperm and egg cells combine during the process of fertilization, the normal two copies of each chromosome type and the normal amount of DNA are restored. Each child therefore receives one of each chromosome type from their mother and one from their father. Which sperm or egg cell gets which copy of each chromosome type is entirely random. In part, this explains why siblings are not genetically identical. It also explains how biological sex is determined. Half of the sperm cells contain an X chromosome, and half contain a Y chromosome. If an egg cell is fertilized by a sperm cell containing an X chromosome, the baby is a girl; if an egg cell is fertilized by a sperm cell containing a Y chromosome, the baby is a boy (Figure 2.17). Y chromosomes are transmitted over multiple generations only in the paternal line, father to son.

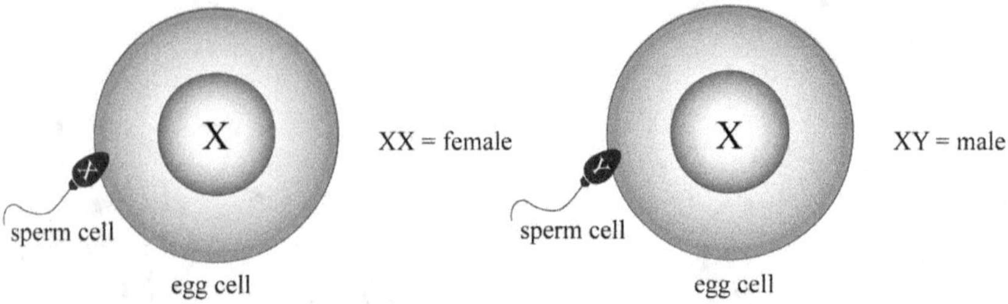

Figure 2.17 Chromosome determination of biological sex

The fundamental concepts of meiosis were discovered in the middle of the 19th century, not by a distinguished professor working at a university, but by a rather unlikely character, a Catholic monk named Gregor Mendel. Mendel made his discoveries through his famous breeding experiments with pea plants in his abbey garden in what is now the Czech Republic (9–10).

Figure 2.18 Gregor Mendel

In addition to the DNA molecules residing in the cell nucleus that code for nearly all our proteins, we also have a relatively small (about 16,500 nucleotide) DNA molecule in our mitochondria. Mitochondria are subcellular organelles that generate chemical energy—they are the cell's power plants. Mitochondria are present in egg cells but not in sperm cells; therefore, only mothers transmit mitochondrial DNA to their children. Since sons cannot transmit mitochondrial DNA to their own children, mitochondrial DNA is inherited over multiple generations solely through the maternal line, mother to daughter.

Genes

Each gene is located at a specific position along one of the chromosomal DNA molecules. A typical human gene is split into segments called exons (Figure 2.19). While a few human genes have only a single exon, the vast majority have multiple exons. The largest has over 300 exons. The DNA segments between exons are called introns. The sum total of all the exons in all of our genes is called the exome. The exome makes up only a small fraction, about 1.5%, of our total genome. The remaining 98.5% of our DNA is either intronic or intergenic (between genes).

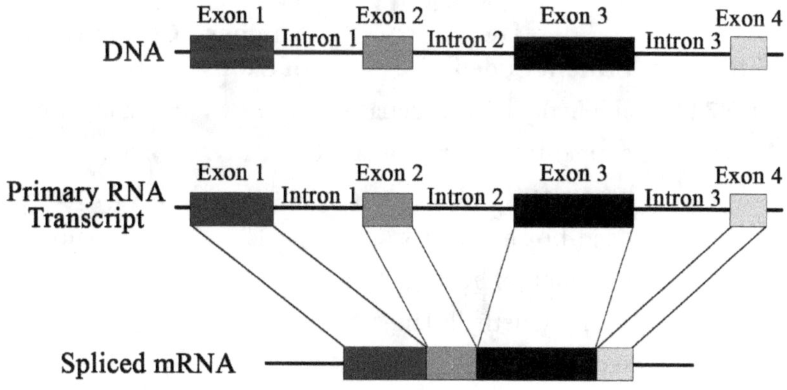

Figure 2.19 Human gene with four exons

The initial (or primary) RNA transcript from a gene contains both exons and introns (Figure 2.19). Before the RNA leaves the nucleus, the introns are removed via a process called splicing. Through splicing and a few other steps, mature mRNA molecules are produced. mRNAs then move from the nucleus to the cytoplasm, where they bind to ribosomes and are decoded to produce protein molecules.

Each of our approximately 20,000 genes is assigned an official name by a committee of the international Human Genome Organization. This avoids the confusion that would ensue if different nations and groups used different names. By convention, official gene names are italicized and, with just a few exceptions, are combinations of capital letters and numbers.

The genes involved in the cases in this book are listed in Table 2.2. The chromosomal positions of the genes are specified by the number of nucleotides in the DNA molecules starting from the tips of the short arms of the chromosomes. A map of the portion of the DNA molecule containing the *HTT* gene from the short arm of chromosome 4 is shown in Figure 2.20. Since DNA polymers are unbranched, chromosomal DNA maps are linear. And because all of our somatic cells have two copies of each chromosome type, we have two copies of each gene (except for the genes located on the sex chromosomes in males).

Table 2.2
Examples of human genes

Gene	Protein Coded	Number of Exons	Chromosome: Position
ALDH7A1	aldehyde dehydrogenase	18	5: 126.6 mb
ATP7B	copper transporter beta	21	13: 52.0 mb
BRCA2	breast cancer 2	26	13: 32.3 mb
DDX3X	dead-box helicase 3	17	X: 41.3 mb
HBB	β subunit of hemoglobin	3	11: 5.2 mb
HMBS	hydroxymethylbilane synthase	14	11: 119.1 mb
HTT	huntingtin	67	4: 3.1 mb
MSH2	mismatch repair protein 2	16	2: 47.4 mb
RYR1	ryanodine receptor 1	106	19: 38.5 mb

mb stands for megabases, or millions of bases (nucleotides). For the purpose of describing positions along a DNA molecule, the terms "bases" and "nucleotides" are used interchangeably.

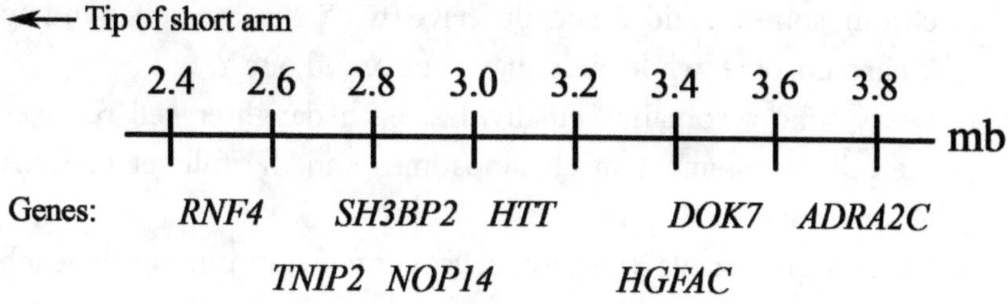

Figure 2.20 Map of a portion of the short arm of chromosome 4

Summary

- Human genetics is the science of the causes and the inheritance of human variation.
- Proteins are the primary functional components of our bodies. We have about 20,000 different proteins, and enormous numbers of copies of each protein type.
- The instructions for the manufacture of our proteins are contained within our DNA molecules.
- These instructions take the form of the *sequence* of adjacent nucleotides along the DNA polymer.
- The genetic code is simple. Three adjacent nucleotides in DNA code for each amino acid in protein. RNA serves as an intermediate in the transfer of sequence information from DNA to protein.
- A gene is a segment of a DNA molecule that codes for a specific protein. Humans have about 20,000 genes, one for each protein.
- Genome is the total complement of DNA within an individual or species.
- The human genome consists of 46 nuclear DNA molecules and one small mitochondrial DNA molecule.
- In our cells, the long, thin DNA molecules are combined together with proteins. These DNA-protein complexes are called chromosomes. The nuclei of our somatic cells have two copies of each of the 23 different types of chromosomes, 46 chromosomes in all.
- The 23 pairs of chromosomes are split between 22 pairs of numbered chromosomes (the autosomes), and the two sex

chromosomes X and Y. Females have two X chromosomes and no Y chromosome, while males have one X and one Y.

- Every time a somatic cell divides, each daughter cell receives the full complement of chromosomes and the full set of DNA instructions.
- Egg and sperm cells (the germ cells) contain only one copy of each chromosome type and half the normal amount of DNA.
- When sperm and egg cells fuse to begin the next generation, the normal number of chromosomes is restored.
- Each person receives one of each chromosome type from their father and one from their mother.
- We have two copies of each gene, except for the genes located on the sex chromosomes in males.
- Each of our approximately 20,000 genes has a specific location along one of the chromosomal DNA molecules.
- Nearly all human genes have multiple coding segments called exons separated along the DNA by introns. Introns are removed from the initial RNA transcripts through a process called splicing.
- Exons comprise only a small fraction, about 1.5%, of our genomes.

Chapter 3

MUTATION

Significance of Mutation

A basic knowledge of mutation is essential for understanding the concepts in this book and genetics in general. Mutation is defined as a change in the nucleotide sequence of a DNA molecule. Mutation provides the raw material for evolution. Without it, species could not evolve, and there would be no life on Earth as we know it.

Many non-geneticists think of mutation as a rare, abnormal event. The word "mutant" often conjures images of individuals with freakish appearance, or severe health problems, or maybe even special abilities (like in the X-Men movies). In reality, however, mutation is a normal and ubiquitous process. Mutations occur in all of our cells throughout our lifetimes. Mutations also occur at each human generation, past and present, as DNA molecules are passed down from parents to children. Therefore, *we are all mutants.*

Mutations lead to what geneticists call DNA sequence variants, or variants for short. As you'll see, variants are responsible for much (but not all) human variation, including differences among us in health and abilities.

Types of Mutations

Many different types of DNA sequence changes occur through mutation. Just about any change in nucleotide sequence you can imagine occurs at least occasionally. A list of the major types of sequence changes is presented in Table 3.1. Mutation types are listed in approximate order of frequency of occurrence in live-born infants, from most to least common.

Table 3.1
Major types of mutation

- Changes in the number of copies of tandem repeats
- Nucleotide substitutions
- Deletions or insertions of nucleotides
- Combined deletions of some nucleotides and insertions of others
- Abnormal numbers of entire chromosomes
- Inversions of segments of nucleotides
- Abnormal joining of DNA molecules from different chromosomes

Let's look in greater detail at some of the more common types of mutations. The most frequent of all occurs at what are called tandem repeats. Tandem repeats are defined as nucleotide sequences that are repeated one after another along a segment of a DNA molecule. The repeated unit can range from one up to many thousands of nucleotides. Roughly 10% of the human genome is composed of tandem repeats (1). Four examples are shown in Figure 3.1: a mononucleotide repeat in the *ZMPSTE24* gene, a compound dinucleotide and mononucleotide repeat in the *CFTR* gene, a trinucleotide repeat in the *HTT* gene, and a

ZMPSTE24: exon 9	CTTTCCTGTG	**TTTTTTTTT**	ATTTGCTGTA
or in abbreviated form	CTTTCCTGTG	**$(T)_9$**	ATTTGCTGTA
CFTR: intron 9	ATTTTTGA	**TGTGTGTGTGTGTGTGTGTGTGTTTTTTT**	AACAGGGA
or	ATTTTTGA	**$(TG)_{11}(T)_7$**	AACAGGGA
HTT: exon 1	CAAGTCCTTC	**$(CAG)_{19}$**	CAACAGCCGC
C9ORF72: intron 1	TCGCGCGCTA	**$(GGGGCC)_3GGGGC$**	GTGGTCGGGG

Figure 3.1 Examples of tandem repeats
Shown are the tandem repeats (bold) within four genes, along with flanking sequences. The subscripts allow an abbreviated display of the repeated sequences.

hexanucleotide repeat in the *C9ORF72* gene. Mutations within each of these tandem repeats can cause disease (see the example of Huntington's disease below).

Tandemly repeated sequences are particularly prone to mutations that either increase or decrease the numbers of repeats (Figure 3.2). The greater the number of adjacent repeats, the higher the rate of mutation (2).

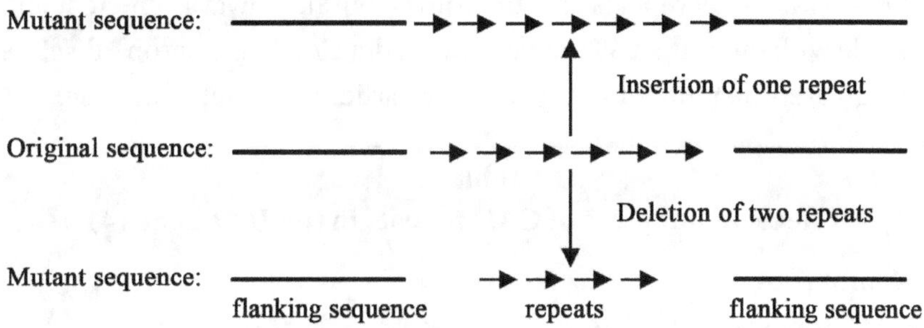

Figure 3.2 Changes in numbers of tandem repeats through mutation

Rarely, variants in tandem repeats are deleterious and cause disease or disability. An example is Huntington's disease (3–4), which is characterized by abnormal movements, personality changes, and gradual loss of cognitive ability. American folksinger Woody Guthrie, Scottish actress Marianna Palka, Australian journalist Sally Pryor, and American scientist Nancy Wexler all developed Huntington's

Figure 3.3 Woody Guthrie

disease. The average age of onset is about 40 years. Today, Huntington's patients die an average of 17 years after onset of symptoms.

Huntington's disease is caused by expansion of the repeating trinucleotide CAG (Figure 3.1) in one of a person's two *HTT* genes. Most *HTT* genes have fewer than 27 tandem CAG repeats (Table 3.2). People who carry a gene with 27–35 repeats are healthy but are at appreciable risk of having affected children because of higher mutation rates. Those with 36–39 repeats are sometimes, but not always, affected, and people with more than 39 repeats always develop Huntington's disease. The greater the number of repeats, the earlier the onset of disease.

Table 3.2
Effects of numbers of CAG repeats in the *HTT* gene (4)

Numbers of CAG Repeats	Effect
less than 27	No effect; normal *HTT* gene
27–35	Healthy, but at risk of having affected children
36–39	Sometimes develop Huntington's disease
more than 39	Always develop Huntington's disease

The second most frequent type of mutation is a nucleotide substitution. In substitutions, the *number* of nucleotides within a segment of a DNA molecule does not change, but the *sequence* of nucleotides does. In the previous chapter, I showed the sequence of nucleotides at the beginning of the gene encoding the beta subunit of the protein hemoglobin (Figure 2.10). This is the sequence in the version of the gene present on the majority of human chromosomes. In Figure 3.4, this sequence is shown again along with two different versions of the gene, each containing a different nucleotide substitution. Because of the redundancy in the genetic code (Figure 2.9), the substitution of a T nucleotide for the C nucleotide at position 9 (counting from the beginning of coding), does not alter the amino acid sequence of the protein. Both gene versions (CAC and CAT) code for histidine (His) at amino acid 3. This variant arose through mutation a very long time ago and is relatively common in human populations. As far as geneticists know, this variant is neutral, having no effect on people's health.

The substitution of a T nucleotide for the A nucleotide at nucleotide position 20, however, is not neutral. This substitution changes the amino acid at position 7 from glutamic acid (Glu) to valine (Val). When present on both chromosomes, this variant causes sickle cell anemia, a disease in which red blood cells adopt an abnormal sickle shape that results in rupture of the cells and severe anemia (5). Within most human populations this variant is very rare, but in Africans, it is relatively frequent, present in up to 30% of chromosomes in some areas. The sickle cell anemia variant is among the most common strongly deleterious variants known.

```
Protein:                                  Met Val His Leu Thr Pro Glu Glu Lys Ser
DNA:      AGCAACCTCA AACAGACACC ATG GTG CAT CTG ACT CCT GAG GAG AAG TCT
                                              ↑ Mutation
DNA:      AGCAACCTCA AACAGACACC ATG GTG CAC CTG ACT CCT GAG GAG AAG TCT
                                                              ↓ Mutation
DNA:      AGCAACCTCA AACAGACACC ATG GTG CAC CTG ACT CCT GTG GAG AAG TCT
Protein:                                  Met Val His Leu Thr Pro Val Glu Lys Ser
```

Figure 3.4 Variants in the *HBB* gene that arose in the past by nucleotide substitution mutation

Another relatively common type of mutation is the deletion or insertion of nucleotides. The deletions or insertions range in size from just a single nucleotide up to millions of nucleotides. The frequency of this type of mutation decreases as the size of the deletion or insertion increases. So, deletions of a single nucleotide are more common than deletions of two nucleotides, which in turn are much more common than deletions of thousands of nucleotides. Deletions or insertions of more than 50 nucleotides are called structural variants.

An example of a variant resulting from a deletion mutation is the deletion of two nucleotides in the breast cancer 1 (*BRCA1*) gene shown in Figure 3.5. Such deletions substantially increase a woman's risk for breast and ovarian cancer. This particular deletion is present in about

Consensus
Protein: Ile Leu **Glu** Cys Pro Ile Cys Leu Glu Leu Ile Lys Glu Pro Val Ser Thr Lys Cys Asp …

Consensus DNA: ATC TTA **GAG** TGT CCC ATC TGT CTG GAG TTG ATC AAG GAA CCT GTC TCC ACA AAG TGT GAC …

Mutation

Variant DNA: ATC TTA GTG TCC CAT CTG TCT GGA GTT GAT CAA GGA ACC TGT CTC CAC AAA GTG TGA CCA …

Variant
Protein: Ile Leu Val Ser His Leu Ser Gly Val Asp Gln Gly Thr Cys Leu His Lys Val **Stop**

Figure 3.5 Deleterious frameshift deletion variant in *BRCA1*

The sequences labeled "Consensus protein" and "Consensus DNA" indicate those sequences present in nearly all human chromosomes.

1% of people with Ashkenazi Jewish ancestry (6). The consensus DNA and protein sequences are those present in nearly all human chromosomes. Note that in this case, the deletion changes the sequence of amino acids downstream of the deletion and creates a new, premature Stop code. This type of variant is called a frameshift. It's a little like when you're typing on a keyboard and accidentally shift your hand over by one column. The result is gibberish.

Sometimes mutation results in an aneuploidy which is the gain or loss of an entire chromosome. Most aneuploidies are incompatible with life and result in a spontaneous (as opposed to elective) termination of pregnancy (Chapter 10). But one such change that is sometimes compatible with life is the gain of a third copy of chromosome 21, or trisomy 21. Trisomy 21 individuals have intellectual disability along with other health problems (7). Shown in Figure 3.6 are the chromosomes from a male with trisomy 21.

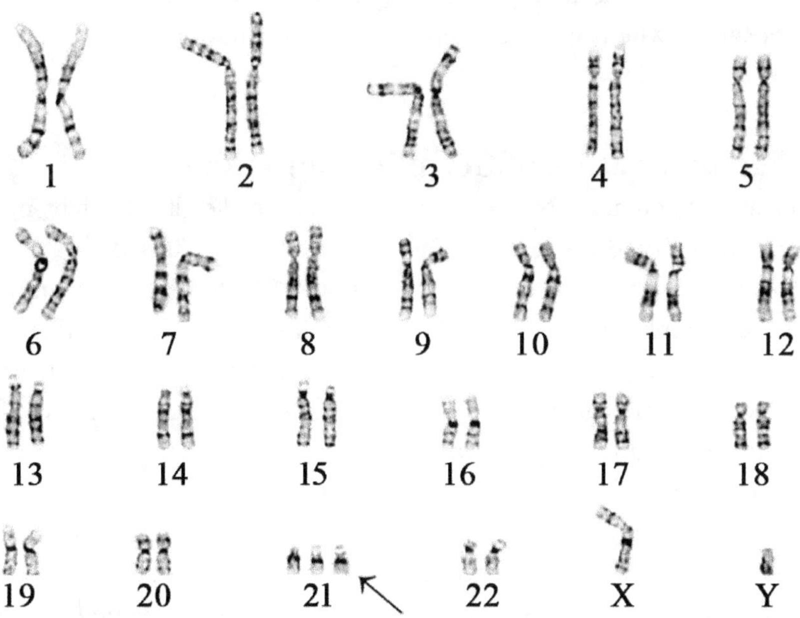

Figure 3.6 Chromosomes from a male with trisomy 21 (from https://thealevelbiologist.co.uk/downs-syndrome-turners-syndrome-and-klinefelters-syndrome). The arrow points to the extra copy of chromosome 21.

In Chapter 2, I described the difference between germ and somatic cells. Mutations that occur in the sperm or egg cells, or the cellular precursors to these cells, may be passed on to the next generation and are called germline mutations. This book deals almost entirely with germline mutations.

However, mutations also occur in all our somatic cells, and at higher rates than in germ cells (8–9). The variants that arise from somatic mutations are not passed on to children, but can greatly impact our health. Cancer, for example, is fundamentally a disease of somatic mutations. The effect of somatic mutations is an active area of research, but geneticists have already learned that somatic mutations are responsible for, or contribute to, not only cancer but several other health problems including kidney cysts, neurological disorders, and aging (10–16). In contrast to variants that arise from germline mutations, which are present in every cell in the body, variants that arise from somatic mutations are present in only a fraction of the cells (17). The fraction depends upon when and where during development or postnatal life the somatic mutation occurs.

Characterization of Germline Mutations

The average numbers of mutations that occur in each live birth by major mutation type are shown in Table 3.3 (2, 18–27). Rates for less common mutation types, like inversions, are thought to be lower than the smallest number in Table 3.3 (28). Some of the numbers in Table 3.3 are strongly supported by data, but the remainder (marked with asterisks) are only current best estimates. These numbers will gradually solidify as research progresses.

Germline mutations occur more often in fathers than mothers (2, 20–21). Germline mutation rates also increase with the age of the parents (19–20, 22, 26). Roughly speaking, mutations occur at random locations along our DNA molecules. No portion of our genomes is immune from mutation. Geneticists currently have no ability to reduce the number of mutations that occur at each generation.

Table 3.3
Average numbers of mutations per birth by type

Mutation Type	Numbers
Changes in number of copies of tandem repeats	1,000*
Nucleotide substitutions	90
Short (<50 nucleotide) deletions or insertions	8
Structural variants (≥50 nucleotide deletions or insertions)	0.2*
Abnormal numbers of entire chromosomes (aneuploidies)	0.003

Asterisks indicate numbers that are only current rough estimates.

A thousand mutations per generation may seem like a lot, but compared to the entire genome, mutations affect only a tiny fraction of our total nucleotides (approximately one mutation for every six million nucleotides). Nearly all of a child's DNA is identical in sequence to the DNA of the parents.

Because the coding regions of our genes, the exons, comprise only about 1.5% of our genomes, and also because exons are depleted in tandem repeats, nearly all mutations occur outside of exons. On average, only about two mutations occur within each child's exome (29–31). Many people are lucky in the genetic lottery of life. The mutations present in their parents' germ cells do not appreciably affect their health and abilities. Many other people, however, are not so lucky. They are born with germline mutations that affect them dramatically. An example is presented in Case 2.

Impact of Mutations

Case 2: Mutation Causes a Disorder

Mei noticed early that her infant daughter, Xiang, was not developing as rapidly as her brother and most other children. Xiang's first simple words didn't come until about 18 months, and at two years she was just beginning to take her first steps. Definitely concerned, Mei brought Xiang to see her pediatrician. The pediatrician confirmed that Xiang was experiencing developmental delay, but couldn't identify a cause. Over the next four years, Xiang was given whole batteries of tests, including extensive blood and urine tests, brain scans, and chromosome analysis. She was also referred to multiple specialists including neurologists, speech therapists, and special educators. The costs mounted, but still, no one could reach a diagnosis.

Although the healthcare providers were stumped, Mei persisted. Finally, her child was referred to a medical geneticist, who ordered genome sequencing tests for Xiang, as well as Mei and her husband. The lab report revealed that Xiang had a deleterious variant in her *DDX3X* gene. The variant was *not* present in Mei or her husband and was therefore the result of mutation. The mutation was a C-to-T substitution that converted a CGA (Gln) code to a TGA (Stop) code. From dozens of other cases reported in the biomedical literature with similar new variants and similar clinical features, it was clear that the variant found in Xiang was the primary cause of her delay.

Mei was happy to finally learn the cause of her daughter's problem. She was greatly relieved to learn that Xiang's condition was not her fault. Still, she wondered why it took so long to reach a diagnosis, and why genome sequencing was not ordered at the start.

Although there is not yet any effective treatment for Xiang's condition, Mei joined a patient support group for families with children who have deleterious variants in *DDX3X*. Through attendance at conferences organized by the patient support group, Mei learned not only about the experiences of other parents with children like Xiang, and the latest management approaches for these children, but also about research to try to find more effective treatments.

She enrolled Xiang in a research trial and raised some money to support the research.

Mei also found through the patient support group and her reading that children like Xiang sometimes develop movement disorders and hearing and vision loss. Therefore, Xiang received physical therapy and regular hearing and vision checks.

The mutation described in Case 2 was deleterious. Deleterious mutations degrade our health and diminish abilities such as vision, hearing, movement, reproduction, thinking, and even socializing. But mutations can also be neutral or even advantageous. Neutral mutations do not appreciably affect our health and abilities; advantageous mutations enhance our health and abilities. Whether a mutation is deleterious, neutral, or advantageous sometimes depends upon the environment. For example, a mutation that imparts resistance to a specific virus may be neutral when the virus is absent, but very advantageous when the virus is present.

In addition, the effects of deleterious and advantageous mutations range broadly in *magnitude*. Some deleterious mutations are lethal (incompatible with life), some cause severe disease and disability, some contribute modestly to disease and disability, and others have only a slight impact on our health and abilities.

Exonic Mutations

Geneticists currently know the most by far about the small fraction (less than 1%) of mutations that occur within or near exons. In general, these "exonic" mutations have a much greater impact on our health and abilities than mutations that occur deep within introns or between genes. The exonic mutations can be broken down into three main groups: synonymous, missense, and loss-of-function.

Synonymous mutations are those that do not alter the amino acid sequence of the coded protein. Many of the three-nucleotide codes in the genetic code are redundant (Figure 2.9). Therefore, many nucleotide substitutions do not change the amino acid sequence of the coded proteins (see the example in Figure 3.4). Synonymous mutations are usually neutral or close to neutral.

Missense mutations are those that result in one or more amino acid substitutions within the coded protein, usually through nucleotide substitutions (Figure 3.4). Missense mutations have a wide range of effects. Since most proteins have segments that tolerate at least some amino acid substitutions, some missense mutations don't appreciably alter the function of the protein. Other missense mutations modestly alter the function of the protein, and some abolish the function of the protein entirely.

Loss-of-function mutations, as the name suggests, result in the total loss of the function of the protein coded by that gene. Loss-of-function mutations include large deletions that remove a significant portion or even all of a gene, and mutations that result in premature Stop codes. An example is shown in Figure 3.5. Although some loss-of-function mutations are tolerated, many are deleterious.

Of the exonic mutations that are not neutral, many more are deleterious than advantageous (32–33). This important conclusion is supported by several lines of evidence. Let's start by comparing the human body to a complex machine with thousands of different, interconnected parts (keeping in mind that no analogy is perfect) (34). The machine has been improved through many rounds of engineering such that it functions with high efficiency. Now, if someone makes a *random* change in the parts—for example modifies the shape of a part, or the connections between two parts, or even removes a part entirely—then the machine probably will not function as well as it did with the original parts. It would be a rare and lucky event for such a random change to improve the machine. So it is with the complex biological machines that are our bodies. Our bodies have evolved over millions of years to function efficiently. When random mutation alters the function of a protein, or eliminates a protein entirely, our bodies usually don't function as well.

Known deleterious exonic variants massively outnumber known advantageous exonic variants. A few million deleterious exonic variants have now been identified in humans, and this number is growing steadily (35). In contrast, only a handful of exonic variants that are clearly advantageous have been discovered (36).

Geneticists also have learned through in vitro (a Latin phrase meaning "outside of living organisms") laboratory experiments that many more exonic mutations are deleterious than advantageous. In one version of such experiments, clever molecular biology tricks are used to mutagenize a gene in a plastic tube, resulting in many forms of the gene with random nucleotide substitutions. Each gene variant is then used to produce protein, and the proteins are tested for function. Experiments like this have now been performed on many different human genes (37–40). Typical results of such experiments for the *BRCA1* gene are shown in Figure 3.7. Synonymous nucleotide substitutions resulted in no change in protein function compared to the unaltered gene. Loss-of-function substitutions (leading to premature Stop codes) resulted in substantial loss of protein function. And missense substitutions resulted in either no change or decreased function ranging from slight to substantial. Few of the missense variants, if any, improved protein function..

Finally, geneticists have now identified *many* thousands of cases in which healthy parents have children with significant disease or disability due to mutation (like in Case 2 above). For instance, deleterious mutations are responsible for about 50% of severe intellectual disability cases (41–42) and about 15% of autism spectrum disorder cases (43–46). Mutations are also responsible for numerous cases of epilepsy, congenital heart disease, muscular dystrophy, skin disease, and many other disorders. In contrast, to my knowledge, no cases have ever been reported in which, for example, nonathletic parents gave birth to an athletically gifted child as a result of a new advantageous mutation.

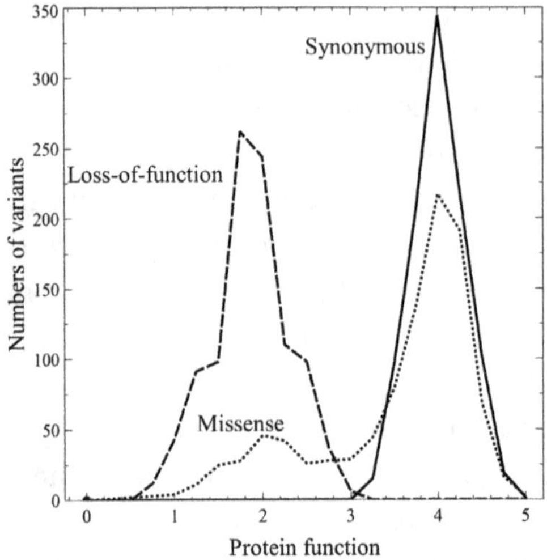

Figure 3.7 In vitro mutagenesis of *BRCA1* (adapted from reference 39, Figure 2d)

Random nucleotide substitutions were introduced into the *BRCA1* gene, and the resulting proteins tested for function. The horizontal axis indicates relative protein function, with the higher numbers indicating greater function. Synonymous variants are represented by the solid curve; missense variants by the dotted curve; and loss-of-function variants by the dashed curve. Variant numbers were normalized.

Extra-Exonic Mutations

In contrast to the small fraction of germline mutations that are exonic and that often have a strong effect on our health and ability, the vast majority of mutations that occur outside of the exons are benign; they have at most only a small effect. If even a handful of the roughly 1,000 mutations that occur each generation were strongly deleterious, humans could not exist.

Geneticists currently do not know what proportion of the abundant extra-exonic mutations are deleterious or advantageous, but it is possible that a larger fraction of these are advantageous compared to exonic mutations (47–49). Extra-exonic mutations do not affect the amino acid sequences of the coded proteins, but they may affect the regulation of genes. Gene regulation involves control of the level and timing of gene activity as well as tissue specificity. Genes may be active at a range of levels from low to high, and may be active only at certain times during our lives. Also, as mentioned in Chapter 2, many of our proteins are manufactured in only some of our tissues.

A trait called lactase persistence results from advantageous variants that alter gene regulation. Lactase is a protein enzyme that catalyzes the breakdown of the sugar lactose. Lactose is present at relatively high concentrations in human breast milk and the milk of other mammals. Infants produce high amounts of lactase so that they can digest the lactose. Most people on the planet stop producing lactase around the time they are weaned. However, in many people from Europe, the Middle East, and a few places in Africa, lactase continues to be produced throughout their lives. This lactase persistence is due to variants near the coding region of the lactase gene that prevent the normal turning off of the gene. The mutations that created these variants occurred many generations ago. People with lactase persistence readily consume dairy products throughout their lives, while those without lactase persistence usually avoid dairy products (50).

Life Course of Variants That Arise Through Mutation

Let's now shift our perspective from individuals to populations. Over about the last century, geneticists have developed an elaborate mathematical framework to describe how the frequencies of deleterious, neutral, and advantageous variants change over time in idealized populations (51–52). This branch of genetics, called population genetics, has been tested many times in real populations, and has generally been found to describe these populations accurately.

Consider a germline mutation involving a T-to-C substitution. Initially, the C variant may only be present in one person in the entire population. This is because at almost all positions along a chromosome, nearly everyone on the planet has two copies of the same nucleotide (Chapter 5). Over time, the frequency of the C variant may increase or decrease in the population.

In the great majority of cases, the new variant, regardless whether deleterious, neutral, or advantageous, persists in the population for only one or a few generations (51–52). Imagine that the T-to-C mutation described in the preceding paragraph occurred in the sperm cell of a woman's father. The woman carries the C variant in each of her somatic cells, but in only half of her egg cells. The probability that she will pass on the C variant to any child is 50%. She may also have no children, or only one or two. Even if she does pass the C variant to at least one child, there is a good chance that her grandchildren will not inherit the variant. The combined effect of these two factors—transmission of only one of the two chromosomes to each child, and limited numbers of children—is that after just a few generations, nearly all new variants vanish from populations simply due to chance.

Table 3.4
Frequencies of neutral variants in populations from population genetics theory

- Mutation is continually adding new variants to populations, thereby increasing the population genetic diversity.
- The vast majority of new neutral variants are lost from populations within a few generations at most.
- New variants that do persist drift randomly up or down in population frequency. On average and over many generations, those variants present at less than 50% frequency tend to vanish from populations, and those at greater than 50% frequency tend to become fixed.
- When there are reductions in population size, called population bottlenecks, variant frequencies change more rapidly than when population size is large.
- If the population bottlenecks are small enough or persist for enough generations, variants move toward either 0% or 100% population frequency and genetic diversity drops.
- Mutations therefore increase the genetic diversity of human populations, and bottlenecks decrease the diversity.

Rarely, a lucky new variant may escape this typical variant loss and increase in a population. The original carrier of the variant may have many surviving children, and many of those children may have inherited the variant. The variant's rise in population frequency is typically a very slow process, occurring over hundreds or thousands of generations and thousands to millions of years. Occasionally, a new variant will even become fixed in the population, so that, using our example, all chromosomes in the population have the C variant and none have the original T variant.

If the variant is neutral, then the changes in population frequency are purely a matter of chance (53). In Table 3.4, I summarize some of the primary population genetics results for neutral variants. While not all of these results are relevant to the current discussion, they will become important later in the book.

Variants That Are Not Neutral

If nearly all new variants with strong effect are deleterious, how can a species survive over long periods of time? Wouldn't these deleterious variants just accumulate, causing individuals in the species to get weaker and weaker, generation after generation, until they could no longer survive to reproduce and the species became extinct? The solution to this mystery, called natural selection, was independently discovered by British biologists Charles Darwin and the lesser-known Alfred Wallace in the middle of the 19th century, about the same time that Mendel was breeding pea plants in his abbey garden. Darwin's book, *On the Origin of Species* (54), was a tremendous breakthrough in the understanding of biology. I highly recommend this book, especially chapters 1–4. It is a masterpiece of careful, detailed observation, marshaled evidence, and brilliant synthesis (55). It certainly ranks among the most important science books ever written.

Darwin and Wallace realized that survival is a great challenge for individuals in the wild. In many species, the majority, often the vast majority, of offspring die before they can reproduce. As just one example, consider a large, mature tree in a forest that produces thousands of seeds each growing season. Only a tiny fraction of those seeds ever grow into another mature tree. If a seed has a deleterious mutation that prevents its roots, stems, or leaves from functioning optimally, then that seed will have an even smaller chance of developing into a new tree, and the new deleterious mutation will likely discontinue. Darwin called this concept "the struggle for existence"; later, others labeled it "survival of the fittest." Natural selection keeps species strong. Without it, no species could survive over long periods of time.

Figure 3.8 Charles Darwin

Geneticists often distinguish negative natural selection (or negative selection) from positive natural selection (or positive selection). Negative selection results in a decrease in population frequency of deleterious variants because, on average, individuals with these variants produce fewer descendants than individuals without these variants. In contrast, positive selection results in an increase in population frequency of advantageous variants because, on average, individuals with these variants produce more descendants than individuals without these variants. One of the best-known examples of positive selection in humans, lactase persistence, was described above. In the past, people who were able to use milk as a food source were more likely to survive and reproduce than those without lactase persistence. Consequently, the regulatory variants that cause lactase persistence gradually increased in populations that raised milk-producing animals like goats and cattle (56).

Another example of a variant that rose to relatively high frequencies in human populations through positive selection is the sickle cell variant described earlier in this chapter (Figure 3.4) (5). This variant is common in people with African ancestry, but is absent in nearly all other individuals. Decades ago, geneticists recognized that the geographic distribution of the sickle cell variant matched the geographic distribution of malaria. People with one copy of the sickle cell variant are more resistant to malaria than people without this variant. Carriers of the variant were therefore more likely to survive infection and reproduce, and its frequency rose in regions where malaria was endemic. The sickle cell variant is an example of a rare type of variant that is advantageous when present on only one chromosome, but deleterious when present on both chromosomes.

In his book, Darwin also clearly distinguished natural from artificial selection. Natural selection is what occurs in nature, and is often a relatively slow process that transpires over many generations. In contrast, artificial selection is what human breeders of pets and agriculturally important species employ. In artificial selection, only those individuals with desirable traits are allowed to reproduce. This results in relatively rapid changes in the frequencies of variants and their resulting traits.

Our modern maize (corn) is an example of artificial selection. Maize was domesticated by Native Americans from the wild grass teosinte, native to Mexico (57–59). Teosinte produces no ears to speak of and is quite different from our current agricultural varieties (Figure 3.9). Similarly, all current breeds of dogs were bred by humans from wolves (60). Every single one of the approximately 400 dog breeds, from Great Danes to beagles to Chihuahuas arose over the last several thousand years through artificial selection (Figure 3.10).

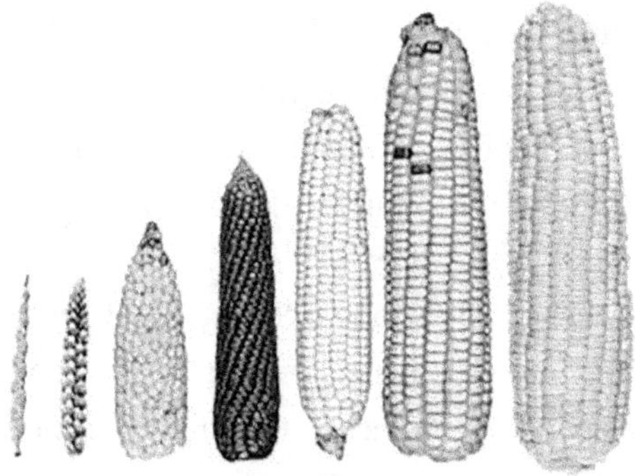

Figure 3.9 Artificial selection of maize (corn)
Teosinte is shown on the far left.

Figure 3.10 Breeds of dogs
The progenitor of all dog breeds, the wolf, is shown at the upper left.

Summary

- Mutation is defined as a change in the nucleotide sequence of a DNA molecule.
- Mutation is a normal and ubiquitous biological process that occurs in all of our cells throughout our lives and between every human generation.
- Approximately 1,000 mutations occur in each human child. We are all therefore mutants.
- Mutation alters nucleotide sequences in many ways, although some types of change, particularly changes in numbers of tandem repeats and nucleotide substitutions, are much more common than other types of change.
- Roughly speaking, mutations occur at random locations within a genome, and therefore the vast majority occur outside of genes.
- New DNA sequences that arise through mutation are called variants. Nearly all new variants have little or no effect on our health

and abilities, but among those with larger effects, deleterious variants greatly outnumber advantageous variants.

- Most new variants vanish from populations within a few generations at most.
- Neutral variants are free to randomly drift up and down in population frequency by chance.
- Natural selection decreases the population frequencies of deleterious variants and increases the population frequencies of advantageous variants.
- Artificial selection is used by human breeders to rapidly change traits in commercial species.
- Mutation has an enormous impact on our species and all other living things.
- Just as we can't prevent hurricanes and earthquakes, we have no ability to stop mutation. Rather, we need to deal with the reality of mutation as best we can.

Chapter 4

INHERITANCE OF CHROMOSOMES, VARIANTS, AND TRAITS

Inheritance of Chromosomes

Through the process of meiosis, each parent contributes one of each chromosome type to each of their children. However, the chromosomes that the parents pass on to their children are not the same as the chromosomes that the parents inherited from their own parents. The simplified version of meiosis that I presented in Chapter 2 (Figure 2.16) omitted an important feature. Prior to the first cell division of meiosis, when each pair of chromosomes aligns, exchange occurs between the DNA molecules in the pair (Figure 4.1). Geneticists call this process recombination. For each chromosome pair, only one or a few exchanges occur in each meiosis. Nearly all of the chunks of grandparental chromosomes inherited by the children are therefore relatively large, harboring tens to hundreds of genes. It may be helpful to compare Figures 2.16 and 4.1.

Figure 4.2 illustrates the inheritance of two chromosome types by a couple's three children. Due to recombination, the children inherit mosaics of the grandparental chromosomes. Like mutation, recombination roughly occurs at random sites along our chromosomes. Because of this, and because it is entirely a matter of chance whether a segment at any point along a chromosome is inherited from the grandfather or

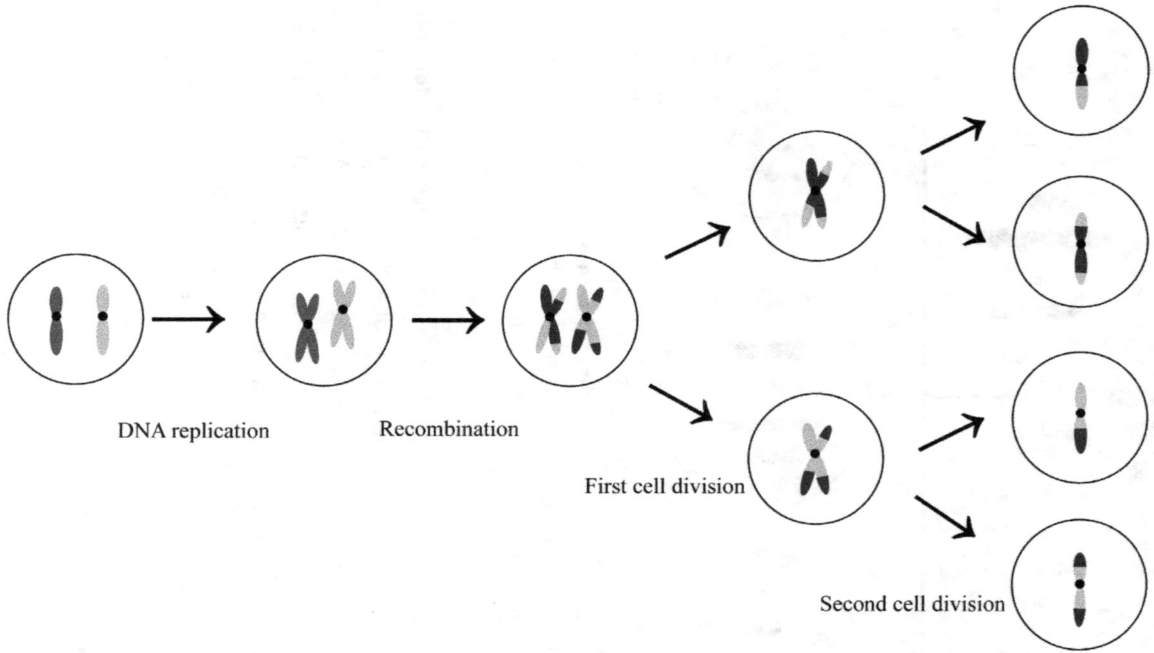

Figure 4.1 Meiotic recombination
Just one of the 23 pairs of chromosomes is shown, but the process is the same in all the other pairs. The chromosome inherited from the mother of the person in whom meiosis takes place is shown in light gray, and the chromosome from the father in dark gray.

grandmother, each child receives a set of chromosomes that is different from their siblings'. Except for identical twins (see below), each child shares half of their DNA sequences with each full sibling.

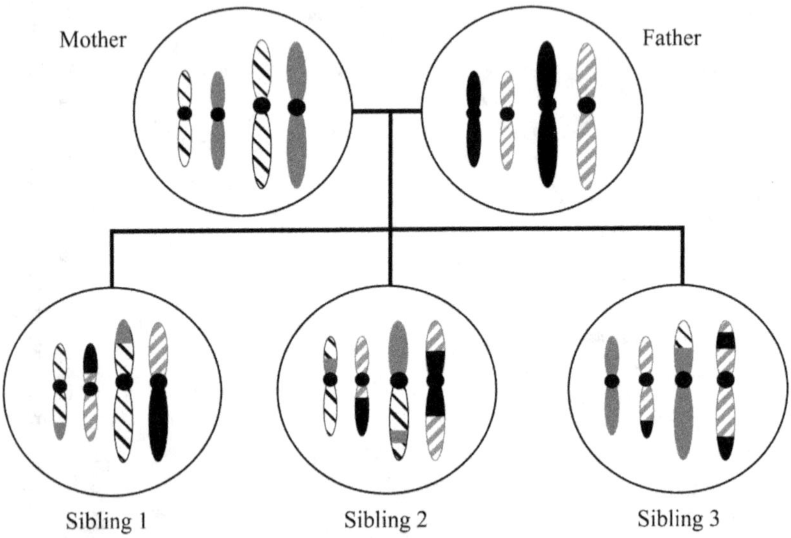

Figure 4.2 Inheritance of chromosomes by full siblings
Just 2 of the 23 pairs of chromosomes are shown. The grandparental origins of the chromosomes are indicated by chromosome fill.

Distinct from mutation, recombination is a process that scrambles existing variants. The analogy of a card game may help you understand the difference between mutation and recombination. Mutation is akin to changing the nature of the cards, say converting an Ace to a Jack. In contrast, recombination is akin to shuffling the deck of cards between hands. Shuffling does not change the nature of the cards, but does change the combination of cards a player receives in each deal.

Inheritance of Variants

Now let's shift focus from chromosomes to individual variants. In clinical laboratories, geneticists detect variants by comparing DNA sequences from a patient to the human reference sequence. With few exceptions, the reference sequence has the consensus (the most common) nucleotide from humanity at each position. Although variants are relatively infrequent along DNA molecules, because we have billions of nucleotides, each person has millions of variants. Nearly all of these variants are inherited from parents; only a small fraction arise from mutation.

When at any specific position along a DNA molecule, the same nucleotide is present on both chromosome copies in a person, geneticists say that this person is homozygous at that position. When two different nucleotides are present, geneticists say that the person is heterozygous at that position. In Figure 4.3 is shown the variant that causes sickle cell anemia, along with some flanking nucleotides. The sickle cell variant occurs exactly at position 5,227,002 on the current reference sequence for the DNA molecule in chromosome 11. Another way to define the position of this variant is that it is the 20th nucleotide from the start of coding (Figure 3.4). Most people have an A nucleotide at this position in both copies of chromosome 11. Such individuals are homozygous A at this position. A minority of individuals, especially those with African ancestry, have an A nucleotide in one chromosome and a T nucleotide in the other, making them heterozygous AT at this position. And a very small fraction of individuals have a T nucleotide in both chromosomes. These individuals are homozygous T and are affected with sickle cell anemia.

	Homozygous A	Heterozygous AT	Homozygous T
Maternal Chromosome 11:	CCTGAGGAG	CCTGAGGAG	CCTGTGGAG
Paternal Chromosome 11:	CCTGAGGAG	CCTGTGGAG	CCTGTGGAG
	↑	↑	↑

Figure 4.3 Homozygosity and heterozygosity
The arrows mark the position of the sickle cell variant.

The squares in Figure 4.4 will help you understand how homozygous and heterozygous variants (the sickle cell variant, in this particular case) are passed from parents to children. For each square, the nucleotides present in one of the mother's egg cells are indicated at the top of each column, and the nucleotides present in one of the father's sperm cells are indicated to the left of each row. Combinations of nucleotides in the cells of a square indicate the two nucleotides present in the two copies of chromosome 11 in a child of the couple. In Figure 4.4A, both the mother and father are homozygous A. Barring an extremely rare mutation event, each egg cell and each sperm cell will have the A nucleotide, and each child will also be homozygous A. In Figure 4.4B, one parent is heterozygous for the A and T nucleotides (the mother, in this case), and the father is homozygous A. Half the cells in this square are AA, and half AT. This means that, on average, half of the children of such a couple will be homozygous A, and half will be heterozygous AT. The results would be exactly the same if the father was heterozygous, and the mother homozygous. In Figure 4.4C, both parents are heterozygous. On average, half the children of such a mating will be heterozygous AT, one-quarter will be homozygous A, and one-quarter (the unlucky ones) will be homozygous T and be affected with sickle cell anemia. Similar squares can also easily be used to predict the outcomes in children when one or both parents are homozygous T.

	A	**A**
A	**AA**	**AA**
A	**AA**	**AA**

A

	A	**T**
A	**AA**	**AT**
A	**AA**	**AT**

B

	A	**T**
A	**AA**	**AT**
T	**AT**	**TT**

C

Figure 4.4 Inheritance of the sickle cell variant
In panel A, both parents are homozygous AA. In panel B, one parent is heterozygous AT and the other homozygous AA. In panel C, both parents are heterozygous AT.

It's important to understand that the results shown in Figure 4.4 are *averages*. Any couple, for example, who are both heterozygous AT, as in Figure 4.4C, and have four children, may have zero, one, two, three, or all four children who are homozygous T and affected. It is a matter of pure chance. The variants that any one child inherits are completely independent of the variants inherited by the other children. Only when the offspring of many such couples are examined, is the fraction that are affected exactly one-quarter.

Heritability

In Chapter 2, I defined human genetics as the science of the causes and the inheritance of human variation. We all know that people vary in many traits. We also know from common experience that children often resemble their parents in these traits. But how can the inheritance of specific traits be described more precisely?

Geneticists use a parameter called heritability to measure the degree of inheritance (1–6). Heritability is defined as the fraction of variability in a disease or trait that is due to variations in inherited DNA. Heritability values range from 0.00 to 1.00, or equivalently, from 0% to 100%. A trait with 0% heritability is not affected at all by inherited DNA sequences, and a trait with 100% heritability is determined solely by inherited DNA sequences.

Heritability can be measured using several different approaches. One obvious approach is to compare a trait between parents and their children. However, because similarities among related individuals are sometimes due not to shared DNA, but rather to shared environment, this is not a perfect way to measure heritability. Consider native language; children nearly always speak the same language as their parents, yet native language has zero heritability.

Fortunately, there are two good ways to disentangle heredity and environment. The first is to study relatives who have been reared and live apart (7). Many variations of this type of study are employed, but one of the most common is to compare children who are adopted as infants to both their biological and adoptive parents. For a highly heritable trait like adult height, adopted children will be more like their biological parents than their adoptive parents. For a trait that is not heritable, like native language, adopted children will be more like their adoptive parents than their biological parents.

The second approach involves twins (7, 8). There are two types of twins: identical twins and fraternal twins. Identical twins (and identical triplets, quadruplets, etc.) arise when an early embryo from the fertilization of one egg with one sperm splits into two (or three or four) groups of cells, and each group of cells develops into a separate baby (Figure 4.5). Identical twins are always same sex (because biological sex is determined by the sex chromosomes), and identical twins share 100% of their DNA, except for somatic mutations. Fraternal twins form when two separate egg cells are fertilized by two separate sperm cells and each resulting embryo develops into a baby. Fraternal twins can be same sex or opposite sex and, like any pair of full siblings, share 50% of their DNA.

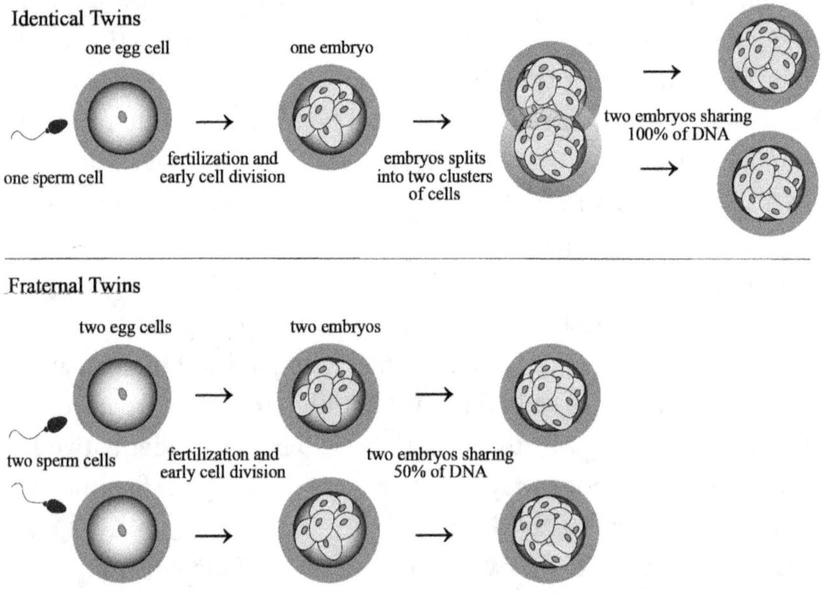

Figure 4.5 Types of twins

For any pair of twins for any trait, researchers can measure the degree of similarity, or concordance, between the pair. Since in childhood twins usually share much of their environment, for instance, same diets and same exposures to infectious organisms, the differences in concordance between sets of identical and fraternal twins for any trait is a measure of the influence of inherited DNA on that trait. Consider facial appearance. Of the three pairs of same-sex twins shown in Figure 4.6, can you tell which are identical and which are fraternal? It turns out that the girls on the left and the boys are fraternal, whereas the girls on

Figure 4.6 Identical and fraternal twins
(photo credit: chintermeyer/Flickr, Creative Commons, CC-BY-SA 2.0)

the right are identical. Full siblings are often similar in appearance, and the fraternal twins in the figure indeed look alike, but they are not exactly the same. Notice, for example, that the boy on the left has a thinner face than his brother. The girls on the right, however, are virtually identical in appearance. That is, of course, why we call them identical twins. Identical twins are surprisingly alike for many traits (9). You may like to view the documentary film *Three Identical Strangers,* which is the story about the meeting of identical triplet boys raised apart from birth.

The concordance for facial appearance is essentially 100% for identical twins, but appreciably less for fraternal twins. This means that facial appearance is a trait with close to 100% heritability. Any trait that does not show 100% concordance between identical twins is not completely heritable and is affected by factors other than inherited DNA.

Heritability is an average measurement obtained by combining results from many people. In any *individual,* however, inherited DNA and nonheritable factors play varying roles. As an example, intelligent parents on average have intelligent children. But as described in Case 2 in Chapter 3, mutations can cause intellectual disability in the children of intelligent parents. Fetal exposure to alcohol during pregnancy can also lead to reduced intelligence in the children of intelligent parents (Chapter 12).

Heritability is difficult to measure. No approach is perfect. Different studies of the same trait usually yield somewhat different values. Nevertheless, the concept of heritability is still useful. Heritability estimates for a number of human disorders and traits are shown in Table 4.1.

Some human traits, like facial appearance and height, have heritabilities at or approaching 100%. Such traits are completely or nearly completely determined by inherited DNA. Many other traits and health problems such as diabetes, intelligence, obesity, cancer, and handedness have intermediate heritabilities. Inherited DNA strongly influences these disorders and traits, but nonheritable factors are also important. And some traits, such as native language or the day of the month on which a person is born, have zero heritability. Our DNA sequences have no appreciable influence on these traits.

Table 4.1
Heritability estimates for selected disorders and traits

Disorder/Trait	Heritability	Disorder/Trait	Heritability
Facial appearance	100%	Obesity	60%
Adult height	85%	Cancer	30%
Juvenile (type 1) diabetes	80%	Handedness	25%
		Native language	0%
Intelligence	60%	Birth day of the month	0%
Adult (type 2) diabetes	60%		

References: height (8, 10), obesity (8, 11), juvenile diabetes (8, 12), adult-onset diabetes (8, 13), intelligence (7, 14–16), cancer (17, 18), and handedness (19, 20).

Geneticists still have much to learn about nonheritable factors, but at the time of writing, I can say that they include environmental exposures, germline and somatic mutations, and random developmental events. Examples abound of environmental exposures that influence our health and abilities; for instance, smoking and lung cancer, obesity and diabetes, and exposure to lead and intelligence. As described in Case 2 from Chapter 3, disease or disability that is completely absent in parents can occur in children through germline mutation. The heritability of cancer is relatively low because cancer is primarily caused by somatic mutation. And random developmental events, such as X chromosome inactivation in females (21), also contribute to human variation.

It is not opinion, but rather *scientific fact,* that most common human health problems and many other human traits have intermediate heritabilities (that is more than 0% but less than 100%). The DNA we inherit from our parents affects our lives greatly, but so do nonheritable factors. So, it's not nature *or* nurture, but rather nature *and* nurture.

Unfortunately, despite this known scientific fact, many people, especially in the past, but even still sometimes today, have entirely denied either the role of inherited DNA or the role of nonheritable factors. More commonly, there are those who, despite paying lip service to the role of both inherited DNA and nonheritable factors, still strongly emphasize one or the other. We all need to be wary of such zealots (22). I think that one important aspect of intellectual strength and courage is not to settle for simple answers to difficult questions, but rather to embrace complexity. I tried in this book to portray heritability realistically and to make clear that in most cases both inherited DNA and nonheritable factors combine to determine traits.

Complex Disorders and Traits

Geneticists often make a distinction between complex and single-gene disorders. Genetically complex disorders and traits are influenced by variants within and near *multiple* genes. The variants that influence complex traits often have substantial population frequencies. Complex disorders and traits are also typically influenced more by nonheritable factors and have lower heritabilities than single-gene disorders. Most common health problems are genetically complex. Examples of genetically complex disorders and traits are shown in Table 4.2.

Table 4.2
Examples of genetically complex disorders and traits

• Hypertension	• Asthma
• Diabetes	• Coronary artery disease
• Inflammatory bowel disease	• Macular degeneration
• Alzheimer's disease	• Obesity
• Height	• Intelligence

Most complex disorders and traits have single-gene subforms. These subforms are caused by variants in single genes, rather than variants in multiple genes. They typically affect only a small fraction of individuals with the disorders and traits. For example, although the great majority of cancer is genetically complex, single-gene subforms, such as Lynch syndrome (colon cancer) and some types of breast cancer, also exist. As another example, although nearly all obesity has complex inheritance, small numbers of individuals have single-gene defects in their satiety systems that cause them not to feel full even after large meals. As a consequence of their constant hunger, they become severely obese (23).

Geneticists currently know much less about complex disorders than single-gene disorders. Still, steady progress is being made, particularly through the development of polygenic indices (Chapter 6).

Single-Gene Disorders and Traits

As the name suggests, each of the thousands of known single-gene disorders and traits is primarily caused by one or two variants in a single gene. Single-gene disorders, also called monogenic or Mendelian (after Gregor Mendel) disorders, are individually uncommon, but since there are a great many of them, they collectively affect large numbers of people (Chapter 11). Single-gene disorders are also usually transmitted between generations with simple patterns, and are caused by highly penetrant variants (more about penetrance shortly).

Examples of single-gene disorders and the genes that are involved are listed in Table 4.3. About 5,000 of our 20,000 genes are currently known to be involved in single-gene disorders, and additional genes involved in single-gene disorders are continually being identified. It is thought that eventually the majority of our genes will be connected to a single-gene disorder (24–25). Geneticists sometimes refer to "disease" genes such as "*BRCA1*, the breast cancer gene," but it's important to understand that all of our genes code for normal, functional components of our bodies and contribute to health. It's only when these genes are altered by mutation that they lead to disease and disability.

Single-gene disorders differ in a number of ways other than the gene involved. As described below, they differ in their patterns of inheritance. In addition, a fraction of genes are involved in multiple disorders. For instance, some deleterious variants in the *RYR1* gene, which codes for an essential skeletal muscle protein, cause congenital muscle disease, while others cause malignant hyperthermia, a severe adverse reaction to anesthetics (Case 5 in Chapter 6). Also, deleterious variants in some genes cause disorders that have different modes of inheritance. Again, using *RYR1* as an example, some forms of congenital

Table 4.3
Examples of single-gene disorders and the genes involved

Disorder/Disability	Gene	Pattern of Inheritance	Estimated Incidence in the US
Cystic fibrosis	*CFTR*	Recessive	1/4,000
Phenylketonuria	*PAH*	Recessive	1/15,000
Duchenne muscular dystrophy	*DMD*	X-linked	1/10,000
Huntington's disease	*HTT*	Dominant	1/20,000
Deafness, prelingual	*GJB2*	Recessive	1/2,000
Deafness, prelingual	*TECTA*	Dominant	1/30,000
Sickle cell anemia	*HBB*	Recessive	1/3,000
Malignant hyperthermia	*RYR1*	Dominant	1/100,000
Congenital myopathy	*RYR1*	Dominant and recessive	1/90,000

muscle disease have dominant inheritance, while others have recessive inheritance. Many single-gene disorders, such as deafness, are caused by deleterious variants in not just one, but multiple genes. Over 200 deafness genes are currently known, and the list is still growing (26). Deafness is therefore not a single disorder, but rather a collection of many single-gene disorders, each caused by abnormalities in a different gene.

Since mutations occur randomly, there are usually many different deleterious variants in each gene involved in single-gene disorders. Unless they are biologically related, different individuals affected with the same single-gene disorder therefore usually carry different deleterious variants. Examples of deleterious variants in the *CFTR* gene that cause the single-gene disorder cystic fibrosis are listed in Table 4.4. About 2,000 deleterious variants in *CFTR* are currently known. Although the details are not important for this book, rules have been developed for naming DNA and protein variants. Both the position of the variant along the gene or protein polymer and the nucleotide or amino acid change are specified (Table 4.4). Note that some deleterious variants change the amino acid sequence of the coded protein while others eliminate protein function altogether by, for example, creating a premature Stop code or altering splicing. Deleterious variants also differ greatly in population frequency. The c.1521_1523del (p.Phe508del) *CFTR* variant is the most common deleterious *CFTR* variant found in people with European ancestry, and is one of the most common deleterious variants of all in that population. But the other variants listed in Table 4.4 have considerably lower population frequencies. With relatively few exceptions, variants that cause single-gene disorders are rare in human populations.

Another important property of single-gene disorders is the set of clinical features exhibited by those affected. Clinical features, also called symptoms or phenotypes, are traits or laboratory results that fall outside normal ranges. There are thousands of recognized clinical features covering all parts of the body and all biochemistry. Examples are abnormal convolutions of the brain surface (lissencephaly), eyes that are farther apart than normal (hypertelorism), unusually fragile bones

Table 4.4
Examples of deleterious variants in the *CFTR* gene that cause cystic fibrosis

DNA Change	Resulting Protein Change	Type	Frequency of Carriers* with European Ancestry
c.1521_1523del	p.Phe508del	Deletion of single amino acid	24 per 1,000
c.350G>A	p.Arg117His	Amino acid substitution	5 per 1,000
c.1652G>A	p.Gly551Asp	Amino acid substitution	0.8 per 1,000
c.1624G>T	p.Gly542*	Premature stop code	0.7 per 1,000
c.3217dup	p.Tyr1073Leufs*3	Frameshift	0.02 per 1,000
c.273+1G>A	p.?	Splicing	0.02 per 1,000
g.117474181-117668151del	Deletion of entire gene	Large deletion	Very rare

From the ClinVar (https://www.ncbi.nlm.nih.gov/clinvar/) and gnomAD (https://gnomad.broadinstitute.org/) databases.

*Carriers are individuals who are heterozygous for the deleterious variants.

(osteopenia), slower than normal infant and child development (developmental delay), and high blood sugar (hyperglycemia).

Clinical features for two single-gene disorders, cystic fibrosis and Huntington's disease, are listed in Table 4.5. Before the advent of clinical DNA testing, single-gene disorders were diagnosed using clinical features alone. Today, in developed nations, they are usually diagnosed through a combination of clinical features and DNA testing. Importantly, patients with single-gene disorders will often exhibit some, but not all, of the clinical features associated with that disorder.

Table 4.5
Major clinical features of cystic fibrosis and Huntington's disease

Cystic Fibrosis	Huntington's
Failure to thrive	Gait ataxia
Elevated sweat chloride	Chorea
Chronic sinusitis	Bradykinesia
Bronchiectasis	Muscle rigidity
Recurrent pneumonia	Neuron loss
Decreased forced expiratory flow (lungs)	Depression
	Cognitive decline
Meconium ileus	Personality changes
Diarrhea	Abnormal eye movements
Male infertility	Weight loss
Pancreatitis	
Clubbing of fingers	

From the OMIM database (https://www.omim.org/)

A comparison of the properties of single-gene and complex disorders and traits is shown in Table 4.6. Splitting health problems into two groups with simple and complex inheritance is convenient and sometimes helpful, but it's also an oversimplification. In reality, single-gene disorders are modified to a degree by secondary genetic variants and by nonheritable factors (see below), and as described above, single-gene subforms of complex disorders are also usually known.

Table 4.6
Comparison of single-gene and complex disorders

Single-gene	Complex
Rare in populations	Common in populations
Caused primarily by variants in one gene	Caused by variants in many genes along with nonheritable factors
Primary causative variants are rare	Causative variants are both rare and common
Primary causative variants have high penetrance	Causative variants mostly have low penetrance

Patterns of Inheritance for Single-Gene Disorders

The main patterns of inheritance for single-gene disorders are autosomal dominant, autosomal recessive, and X-linked. Examples of disorders with each type of inheritance are shown in Table 4.3. In autosomal dominant single-gene disorders, affected individuals have one functional copy of a gene and one dysfunctional copy. The dysfunctional copy of the gene can be inherited from either parent or can arise through mutation. When inherited, each child has a 50% chance of receiving the dysfunctional copy of the gene (Figure 4.4B). Autosomal dominant inheritance is a little like a bicycle—when *either* the front or back tire is flat, you can't ride.

Shown in Figure 4.7 is a four-generation tree of a family affected with a dominant disorder. By convention in these types of drawings, males are represented by squares and females by circles. Each generation is in a separate row. So, two great-grandparents are shown in the top row, then the grandparents in the second row, parents in the third, and children in the fourth. Affected individuals are indicated by filled symbols. Note that about half of the offspring of affected individuals in the family are also affected. None of the offspring of unaffected individuals are affected.

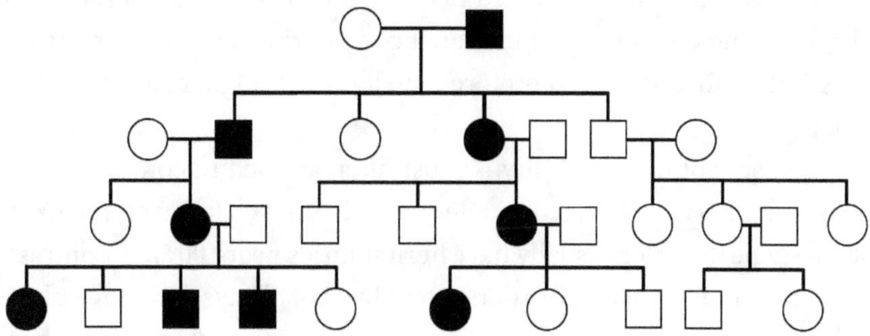

Figure 4.7 Four-generation family tree with dominant inheritance

Females are represented by circles; males by squares. Each generation is on a separate row. Filled symbols indicate affected individuals.

In autosomal-recessive single-gene disorders, individuals with one functional copy and one dysfunctional copy of a gene are unaffected. Only individuals in which both copies of a gene are dysfunctional are affected. Affected children arise when both the mother and the father carry a deleterious variant in the same gene (Figure 4.4C). The deleterious variant can be the same or different in the two parents. Since the parents are unaffected, they usually do not know that they carry the deleterious variants. Also, for recessive disorders, the numbers of people in populations who are heterozygous carriers of deleterious variants greatly outnumber affected individuals. For instance, about 80 out of every 1,000 African Americans carries the deleterious variant for sickle cell anemia, but only about 1.5 out of every 1,000 African Americans are affected with the disease. Autosomal-recessive inheritance is a little like vision. When a person loses vision in one eye, they can still see, but when a person loses vision in both eyes, they are blind.

In X-linked single-gene disorders, the gene involved is located on the X chromosome. Typically, mothers have one functional copy and one dysfunctional copy of an X chromosome gene and are healthy. Their daughters, regardless of which gene they inherit from their mothers, are also usually unaffected because their fathers rarely carry a dysfunctional copy of the gene. However, half the sons of carrier mothers, on average, will be affected because they do not have a second X chromosome containing a functional copy of the gene. The frequency of X-linked disorders is therefore usually much higher in males than females.

The concept of heritability is most often applied to complex disorders and traits rather than to single-gene disorders. However, recessive single-gene disorders usually have heritabilities near 100%. In contrast, dominant and X-linked disorders often, but not always, have lower heritabilities because they can arise via mutation rather than being inherited. For many dominant disorders, new mutation is responsible for a substantial fraction, or even all, of the affected individuals.

Penetrance and Variable Expressivity

Penetrance and variable expressivity are two important properties of deleterious variants. Penetrance is defined as the fraction of individuals with one or more *specific* deleterious variants who are affected with a disorder or who have a particular trait. Penetrance is usually reported on a scale of 0% to 100%. Incomplete penetrance (less than 100%) for deleterious variants is common in single-gene disorders, and nearly universal for complex disorders. As an example of incomplete penetrance, not all women who have a deleterious variant in one of their breast cancer genes develop cancer. Only about 65% of women who have a deleterious loss-of-function variant in one of their *BRCA1* genes develop cancer by the age of 70, and many never develop cancer during their lives (27–28).

Another well-studied example is the E4 variant within the *APOE* gene. This variant is the strongest common genetic risk factor currently known for the development of late-onset Alzheimer's disease. Individuals who are heterozygous or homozygous for E4 have 3 and 12 times the risk, respectively, for developing the disease compared to people without E4 (29–30). However, the penetrance for developing Alzheimer's by age 75 is only about 15% for those who are heterozygous E4, and 30% for those who are homozygous E4. About 60% of individuals who are homozygous E4 will never develop Alzheimer's during their lifetimes. In addition, about 40% of individuals with Alzheimer's *do not* have E4. Geneticists therefore cannot, by testing *APOE* alone, determine whether a person will or will not develop Alzheimer's disease.

Variable expressivity means that individuals such as a pair of siblings who carry exactly the same deleterious variant (or variants, in the case of recessive disease) often have different disease courses and/or severities and also often have different sets of clinical features. At least part of the explanation for variable expressivity is that variants in other genes modify the effect of the primary deleterious variant(s), and also that different individuals experience different nonheritable factors.

Combined Effect of Risk Factors

Although variants with the strongest effects and the highest penetrance play a large role in disease and disability, it would be a mistake to ignore the combined effect of the many variants with weaker effects and lower penetrance. Geneticists have now completed many studies comparing DNA from a group of individuals with a specific complex disorder or trait (the cases) with DNA from a matched group of individuals without the disorder or trait (the controls) (31-32). Through such case/control studies, it has been found that tens, hundreds, or even thousands of different variants, mostly located outside of exons, have an effect upon individual complex disorders and traits.

In general, as the penetrance of a variant increases, the population frequency of the variant decreases. Single-gene disorders are nearly always caused by highly penetrant variants that are rare to very rare in populations. Genetically complex disorders and traits will typically have a few risk variants with high penetrance and very low population frequencies (the single-gene subforms), some risk variants with medium penetrance and low frequencies, and many risk variants with low penetrance and moderate to high frequencies.

To help you understand the impact of numerous variants and other risk factors on an individual's health, I've prepared the "arrow plots" shown in Figures 4.8 and 4.9. Although these plots are not a perfect representation of what occurs in nature, they are still a useful learning tool. They may seem a little complicated at first, but are not really that difficult to comprehend. Importantly, the plots portray risks for specific health problems in *individuals*, not populations.

Each of the different risk factors, which may be either a DNA variant or a nonheritable factor, is represented in Figures 4.8 and 4.9 as a separate arrow. An arrow pointing up represents a factor that increases risk for the disease, and an arrow pointing down one that decreases risk. The length of each arrow indicates the magnitude of risk; long arrows represent factors with high risk, and short arrows factors with low risk. Short arrows are common; long arrows are rare. For many disorders there are hundreds or even thousands of arrows/risk factors. For practicality, only a small fraction of the arrows are shown in the plots.

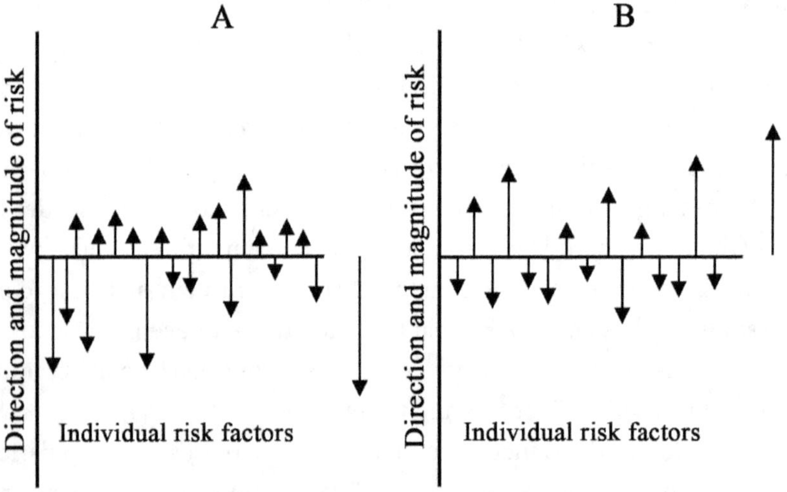

Figure 4.8 Risk factor plots
Panel A: plot for a person with decreased overall risk for disease. Panel B: plot for a person with increased overall risk.

To determine a person's overall risk for developing a disorder, the *lengths* of all the up-pointing arrows are added together, and the *lengths* of all the down-pointing arrows are also added together. The total length of all the down arrows is then subtracted from the total length of all the up arrows. This gives a cumulative arrow (shown at the far right in each plot) that represents a person's overall risk for disease. If this cumulative arrow points up, then the person is at higher risk than the average person. If the cumulative arrow points down, then the person is at lower risk.

The arrow plot for a person who has a cumulative *decreased* risk of developing a complex disorder compared to the average person is shown in Figure 4.8A. The down-pointing arrows outweigh the up-pointing arrows. The plot for an individual who has a cumulative *increased* risk for a complex disorder is shown in Figure 4.8B. Here the arrows pointing up are predominant. The plots help us understand why knowing any single risk factor for a complex disorder, even if it is a relatively strong one, like the *APOE* E4 variant, is insufficient to predict the overall risk for an individual.

The arrow plots can also help us understand single-gene disorders. In single-gene disorders, there is a primary deleterious variant (or variants, in the case of recessive disease) that outweighs all other risk factors (see the long up-pointing arrows in Figure 4.9). However, just as in complex disease, weaker risk factors do exist (33–38). Consider two siblings who share the exact same deleterious variant for a dominant single-gene disorder. The sibling in Figure 4.9A has mostly up-pointing arrows for the other risk factors, while the sibling in 4.9B has mostly down-pointing arrows for the other risk factors. The sibling in 4.9A will therefore likely be more severely affected than the sibling in 4.9B. This helps explain variable expressivity. The clinical features for the sibling in 4.9B may even be so mild that this person is not considered to be affected at all (39). This helps explain incomplete penetrance.

The arrow plots also nicely portray how variable traits like blood pressure, adult height, and intelligence are determined. Trait values are plotted on the vertical axis, with the horizontal axis indicating the average value for the trait. The arrows then indicate variants or nonheritable factors that individually either add to (up arrows) or detract from (down arrows) the average value for the trait. Cumulative up arrows indicate individuals who are above average for the trait, and cumulative down arrows indicate individuals who are below average.

Aspects of the *inheritance* of disorders and traits can also be gleaned from the arrow plots. When both parents have a complex disorder like type 2 diabetes or are both above average for a trait like height, then their children will also tend to be at high risk for diabetes or above average in height. This is because the children will inherit more factors that increase risk for the disease or contribute to height than factors that

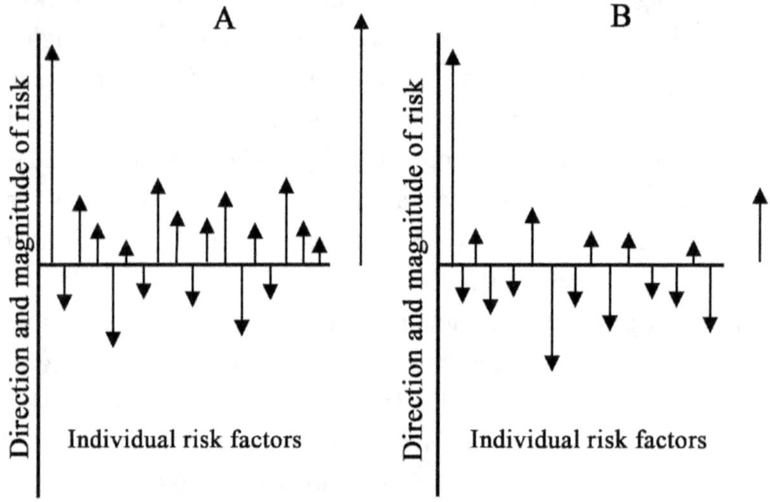

Figure 4.9 Risk factor plots for a pair of siblings with the same dominant deleterious variant

Panel A: Plot for a sibling with a dominant deleterious variant, but with most other factors increasing risk. Panel B: Plot for a sibling with the same dominant deleterious variant as in Panel A, but with most other factors decreasing risk.

reduce disease risk or detract from height. Due to the chance inheritance of variants and also to variable nonheritable factors, the children will not always become affected or match their parents closely in height.

In Chapter 6, risk indices, which are similar to the cumulative arrows in the arrow plots, will be introduced. Risk indices are becoming important in healthcare generally, and in disease prevention in particular.

Summary

- Parents randomly transmit one of their two genes to each of their children. This process is independent for each child.
- Independent segregation of genes and meiotic recombination increase the diversity of genomes.

- Geneticists measure the degree of inheritance of a disorder or trait using a parameter called heritability. Heritability is often estimated through twin or adoption studies.
- Many common health problems and other human traits are strongly influenced by the DNA we inherit from our parents, but are also strongly influenced by nonheritable factors such as environmental exposures and mutation. Hence, they have heritabilities substantially greater than 0%, but less than 100%.
- Genetically complex disorders are influenced by variants in many different genes and by nonheritable factors.
- Single-gene disorders are predominantly caused by one or two rare, highly penetrant deleterious variants in a single gene. Unrelated individuals affected with the same single-gene disorder usually have different deleterious variants in that gene.
- Single-gene disorders are primarily inherited in simple dominant, recessive, and X-linked patterns.
- The effects of individual risk factors, both heritable and nonheritable, can be combined to estimate a person's overall risk for disease.

Chapter 5

HUMAN EVOLUTIONARY HISTORY AND GENETIC DIVERSITY

Human Evolutionary History

I'll turn now to the fascinating and important topic of human evolutionary history. Knowledge of our evolutionary history is important for understanding ourselves, other living things, and how genetics affects our health and lives.

A few key dates in our evolutionary history are listed in Table 5.1 (1–2). Our planet is old, about 4.5 billion years old. Simple living organisms are also very old; the first known unicellular organisms date to sometime around 3.7 billion years ago. Gradually, over very long periods of time, the first relatively simple life forms evolved into more complex organisms. Since scientists are still learning a great deal about the history of life on our planet, I suspect that the dates in Table 5.1 will be revised in future. Still, they offer important perspective.

From the numbers in Table 5.1, you can see that humans are a very young species. Anatomically modern humans have been around only about 200,000 years, or 0.00005 (five–one hundred thousandths) of the time that life has existed on our planet. In comparison, although the species were undoubtedly different from those that exist today, animals that we would all recognize as fish have been around for about

530 million years, dragonflies for 320 million years, and snakes for 130 million years. Snakes are therefore about 650 times older than people.

Table 5.1
Selected key dates in human evolutionary history

Event	Years Ago
Formation of Earth	4.5 billion
First living organisms	3.7 billion
First animals	600 million
Earliest land plants	470 million
First mammals	230 million
Demise of the dinosaurs	66 million
Last common human-chimp ancestor	7 million
Appearance of Neanderthals	600,000
First modern humans	200,000
Modern human departure from Africa	60,000
Extinction of Neanderthals	40,000
Advent of agriculture	12,000

Biologists often represent the evolution of life on our planet as a tree, with branches leading to different groups of organisms. At the far ends of the branches are the individual species living today. Species that are closely related to each other, such as different species of ducks, are found on close, adjacent branches that fork near the ends of the branches. Species that are not closely related, such as dogs and bacteria, are located far apart on the tree, with branches that fork close to the trunk. Many trees of life have been produced; they can easily be found by searching the internet for "evolutionary trees of life." One such tree is shown in Figure 5.1.

It's useful to have a basic understanding of how evolutionary trees are generated. In the past, species were grouped by appearance, anatomy, and biology. Today, species are grouped largely by sequencing their genomes (3). The more closely related are two species, the more similar are their DNA sequences. The closest relative to any particular species is the species with the most similar DNA sequence.

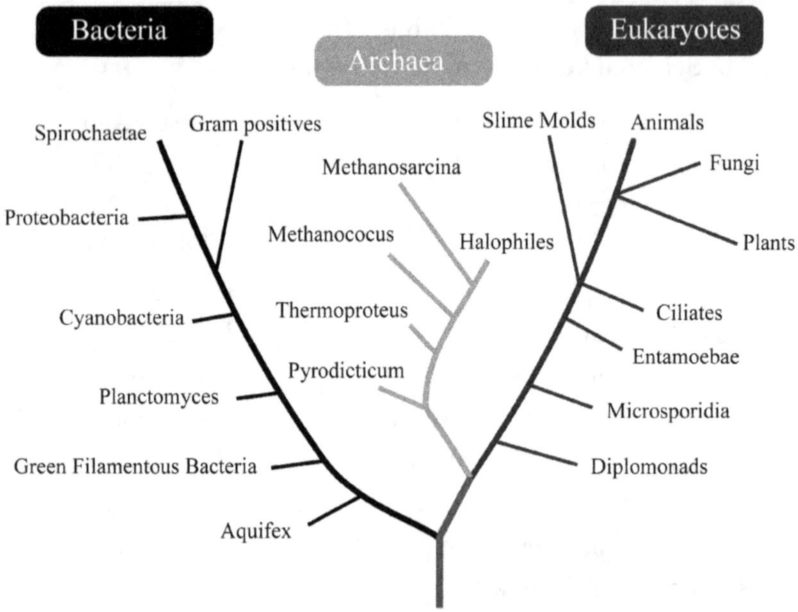

Figure 5.1 Evolutionary tree of life

Like real trees, the tree of life has many branches that are dead and do not extend to the ends of living branches. These dead-end branches represent extinct species. From the fossil record, it is known that vast numbers of species that once lived are now extinct.

DNA sequencing has confirmed that our closest living relatives are the great apes: orangutans, gorillas, and chimpanzees (4–5). Orangutans are now native only to Southeast Asia. Gorillas and chimps are native only to Africa. The portion of the tree of life that contains the great apes and humans is shown in Figure 5.2.

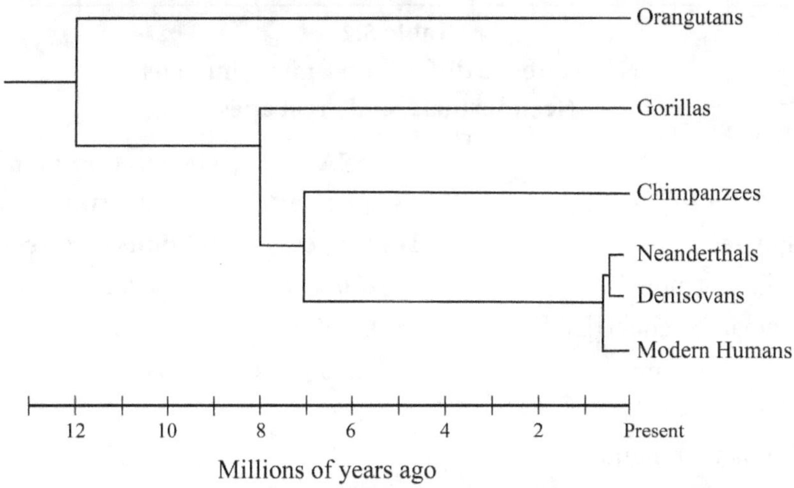

Figure 5.2 Evolutionary tree of great apes and humans
Before they became extinct, Neanderthals and Denisovans contributed DNA to modern humans.

Of the great apes, chimpanzees are closest in DNA sequence to ourselves (Figure 5.3). The common ancestor of humans and chimps lived roughly seven million years ago. Humans and chimps have DNA sequences that are approximately 98.5% identical (Table 5.2). Gorillas are almost as close to us as chimps, and orangutans only a little more distantly related.

Figure 5.3 Chimpanzees are our closest living relatives

Table 5.2
DNA sequence differences among humans, Neanderthals, and great apes

Species	DNA Sequence Differences	Time to Common Ancestor (millions of years)
Human-Human	0.10%	0.2
Human-Neanderthal	0.15%	0.6
Human-Chimp	1.5%	7
Human-Gorilla	1.8%	8
Human-Orangutan	3.4%	12

Over the approximately seven million years between the common human-chimp ancestor and the present, our ancestors changed considerably in appearance (Figures 5.3 and 5.4), behavior, and abilities (6–11). Their arms shortened and their legs lengthened. They learned to make stone tools and to use fire. They gained the ability to speak. Their brains increased in size about threefold.

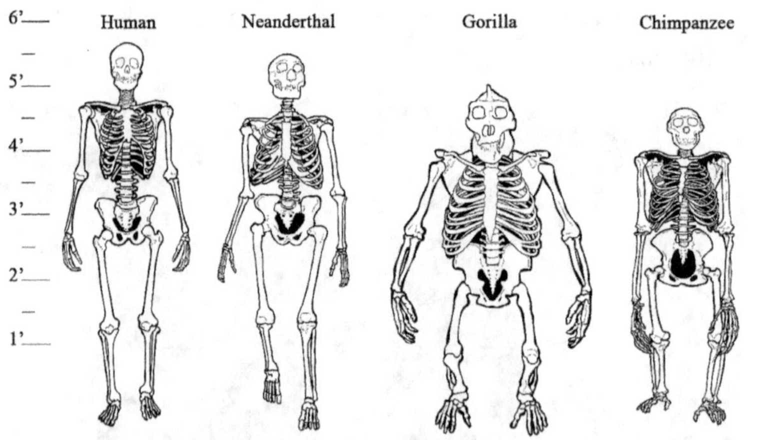

Figure 5.4 Comparison of human, Neanderthal, and ape skeletons

Along the way from the common human-chimp ancestor to anatomically modern humans, there were at least several—and probably many—dead ends. These are species who once lived and were more closely related to us than to chimps, but did not survive to the present. Scientists know about these dead ends from the fossil record which, although sparse, still presents a panoply of extinct relatives (6).

The closest of these extinct relatives to ourselves, the Neanderthals, merit special attention. Neanderthals split from the ancestors of modern humans about 600,000 years ago (Figure 5.2). At their peak, Neanderthals and a second group closely related to Neanderthals, called Denisovans, occupied much of Europe and Asia (10–11). Neanderthal skeletons are distinct from modern human skeletons but still quite close (Figure 5.4). Like our ancestors, Neanderthals used stone tools and hunted. They may also have used fire, buried their dead, worn jewelry, built boats, and created artwork. Although only their bones and none of their flesh remain, artists have used the skeletons to attempt to recreate the appearance of Neanderthals. Two of these attempts are shown in Figure 5.5; many others can easily be found on the internet. Neanderthals looked pretty much like ourselves.

Figure 5.5 Re-creations of Neanderthals

From the fossil record and from DNA sequencing in living (extant) humans, geneticists have learned that anatomically modern humans originated in Africa roughly 200,000 years ago (11–14). The oldest modern human fossils come from Africa. Although *all* humans are very close genetically, when trees of extant humans are constructed, the branches that split off earliest are those leading to Sub-Saharan

Africans. All extant human populations living outside of Africa, including Europeans, Asians, Native Americans, Aboriginal Australians, and Polynesians, are more closely related to each other than they are to Sub-Saharan Africans.

Although anatomically modern humans apparently made a number of earlier forays out of Africa (11), the main thrust of modern human emigration from Africa started about 60,000 years ago (11–14). Modern humans then spread throughout the entire world (Figure 5.6). This colonization process was still ongoing in historic times. Polynesians in their outrigger boats were colonizing islands in the Pacific as recently as 700 years ago. The emigration of modern humans out of Africa was certainly not as simple as indicated in Figure 5.6. It undoubtedly involved multiple forward migrations, backward migrations, extinction of populations, and splitting and mixing of populations (15–16). I won't attempt to describe current detailed knowledge in this area because it is not essential for the purposes of this book, and because this field is advancing so rapidly that any such attempt would become outdated in just a few years.

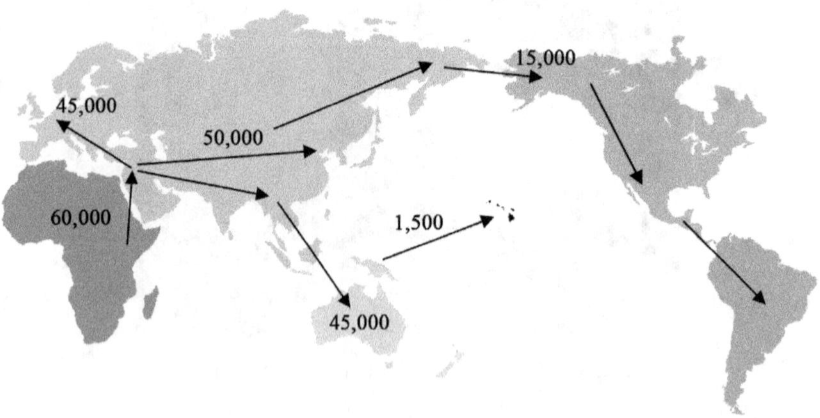

Figure 5.6 Colonization of the world by modern humans coming out of Africa

The numbers accompanying the arrows indicate approximately how many years ago the migrations occurred.

Along their way out of Africa, our modern human ancestors encountered Neanderthals. Neanderthals became extinct about 40,000 years ago, relatively soon after contact with modern humans. Since Neanderthals lived successfully in Europe and Asia for hundreds of thousands of years before meeting humans, it's a good bet that our ancestors were at least a major contributor to their extinction.

A breakthrough in the study of human evolution came in 1985, when Svante Pääbo reported that ancient DNA extracted from the bones and other tissues of long-dead individuals, although degraded, could still be sequenced (17). Since then, methods for sequencing and analyzing ancient DNA have improved markedly (15, 18–19). DNA has now been sequenced from over 100,000 human and human-like remains ranging in age from the near present to about 430,000 years ago (20). Sequencing of ancient DNA has revolutionized understanding of human evolutionary history. In 2022, Pääbo won a Nobel Prize for his work.

Neanderthals are now extinct, but in a way, they live on through us. In 2010, Pääbo's research group reported the genome sequencing of Neanderthal DNA (21). Through analysis of these sequences, geneticists learned that some of our ancestors reproduced together with Neanderthals, with the result that all or nearly all people living today, except those from Sub-Saharan Africa, have at least 1%–2% Neanderthal DNA in their genomes (21–24). People from Asia and Oceania, in addition to their Neanderthal ancestry, also have Denisovan ancestry (25). Sub-Saharan Africans do not have Neanderthal DNA because Neanderthals did not live in Africa and because the ancestors of Sub-Saharan Africans remained in Africa when other people emigrated to the rest of the world. The absence of Neanderthal DNA in the genomes of Sub-Saharan Africans is further evidence for the African origin of modern humans.

In people with "Out-of-Africa" ancestry, the Neanderthal and Denisovan DNA has been broken up into small segments through meiotic recombination. The children of human-Neanderthal matings had one human chromosome and one Neanderthal chromosome of each type (Figure 5.7). But when these children reproduced with humans with little or no Neanderthal ancestry, the human- and Neanderthal-containing chromosomes recombined. Eventually, after many generations, the remaining Neanderthal DNA was reduced to small segments.

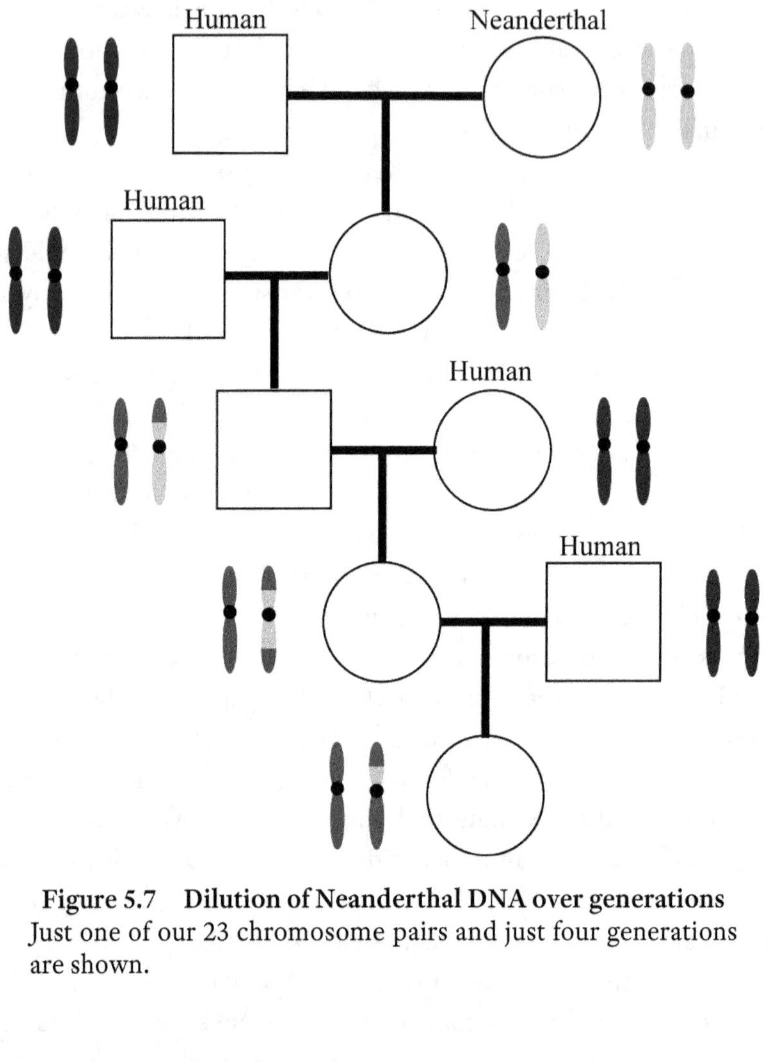

Figure 5.7 Dilution of Neanderthal DNA over generations
Just one of our 23 chromosome pairs and just four generations are shown.

Human Genetic Diversity

Mutation guarantees that individuals within a species do not have identical genomes. However, the differences in DNA sequences among individuals of any species are much smaller than the differences between species. Humans differ from one another in DNA sequence by about 0.1%, but humans and chimps differ by about 1.5%, 15 times the level of human-human differences (Table 5.2).

All people on our planet are very similar genetically. It doesn't matter if the person is from New Guinea, Japan, the Congo, or Italy. If the

DNA from any two people on the planet is sequenced and aligned, the sequences are found to be 99.9% identical; they differ only in about 1 out of every 1,000 nucleotides (26–27). The most common biological definition of "species" is a group of organisms that can reproduce together and produce fertile offspring. Since all matings of humans from anywhere on the planet result in healthy, fertile children, *all* people clearly belong to one common species.

Even though all human genomes are very similar, because our genomes are so large (6.2 billion nucleotides), there are millions of differences. Comparison of genomes from any two unrelated people reveals roughly five million differences in sequence. To a considerable degree, these differences are responsible for the variation among us.

An example of nucleotide sequences from different human DNA molecules is shown in Figure 5.8. This small DNA segment is from the beginning of the gene coding for the beta subunit of the protein hemoglobin (the *HBB* gene) (see also Figures 2.9, 3.4, and 4.3). Regardless whether the DNA sequences are from different people or from the two copies of the gene in any single individual, the sequences are nearly identical; only rarely do they differ.

AGCAACCTCA AACAGACACC ATG GTG CAC CTG ACT CCT GAG GAG AAG TCT
AGCAACCTCA AACAGACACC ATG GTG CAC CTG ACT CCT GAG GAG AAG TCT
AGCAACCTCA AACAGACACC ATG GTG CA**T** CTG ACT CCT GAG GAG AAG TCT
AGCAACCTCA AACAGACACC ATG GTG CA**T** CTG ACT CCT GAG GAG AAG TCT
AGCAACCTCA AACAGACACC ATG GTG CAC CTG ACT CCT GAG GAG AAG TCT
AGCAACCTCA AACAGACACC ATG GTG CAC CTG ACT CCT GAG GAG AAG TCT
AGCAACCTCA AACAGACACC ATG GTG CAC CTG ACT CCT G**T**G GAG AAG TCT
AGCAACCTCA AACAGACACC ATG GTG CA**T** CTG ACT CCT GAG GAG AAG TCT
AGCAACCTCA AACAGACACC ATG GTG CAC CTG ACT CCT GAG GAG AAG TCT
AGCAACCTCA AACAGACACC ATG GTG CA**T** CTG ACT CCT GAG GAG AAG TCT
AGCAACCTCA AACAGACACC ATG GTG CA**T** CTG ACT CCT GAG GAG AAG TCT
AGCAACCTCA AACAGACACC ATG GTG CAC CTG ACT CCT GAG GAG AAG TCT
AGCAACCTCA AACAGACACC ATG GTG CAC CTG ACT CCT GAG GAG AAG TCT
AGCAACCTCA AACAGACACC ATG GTG CA**T** CTG ACT CCT GAG GAG AAG TCT
AGCAACCTCA AACAGACACC ATG GTG CAC CTG ACT CCT GAG GAG AAG TCT
Met Val His Leu Thr Pro Glu Glu Lys Ser
His Val

Figure 5.8 Sequences from the beginning of the *HBB* gene from different individuals

The left arrow in Figure 5.8 marks a position (nucleotide 9) in the *HBB* gene at which variation is common. Nearly all human DNA molecules have either a T or a C nucleotide at this position, and this substitution does not alter the amino acid sequence of the coded protein (Figure 3.4). Frequencies of the T and C variants in a number of human populations are shown in Table 5.3. Common sequence variants such as this one mostly arose many thousands of generations ago, and are almost always neutral or close to neutral.

The right arrow in Figure 5.8 (at nucleotide 20) marks a rare variant, the sickle cell anemia variant. Although this variant is relatively common in people with African ancestry, it is rare in other human populations (Table 5.3).

Table 5.3
Frequencies of variants at nucleotides 9 and 20 in the *HBB* gene in selected human populations

	Nucleotide 9 Variant		Nucleotide 20 Variant	
Population	**T**	**C**	**A**	**T**
East Asian	50%	50%	100%	0%
South Asian	35%	65%	100%	0%
Latin American	34%	66%	100%	0%
European	16%	84%	100%	0%
Ashkenazi Jewish	14%	86%	100%	0%
African American	13%	87%	96%	4%

From the gnomAD database (https://gnomad.broadinstitute.org/)

Deleterious Variants Present in People

Nearly all people carry one or more highly penetrant deleterious variants, mostly for recessive disorders. These are variants that geneticists would categorize as causative for a single-gene disorder or that would be lethal in the homozygous state. Estimates for the average number of such variants per person generally range from 3 to 10 (28–30). However, these numbers depend upon the definitions of "highly penetrant" and even "disorder." They are also likely to be at least somewhat underestimated since not all deleterious variants can be easily detected with current sequencing technologies and since many deleterious variants have not yet been identified.

People also carry *many* deleterious variants of lower penetrance. Because negative natural selection is not as efficient at removing these lower-penetrance variants, all people have some deleterious variants with intermediate penetrance and thousands of deleterious variants with low penetrance. There is therefore no such thing as a genetically perfect person. Even the most fabulous athletes, the richest business people, the most brilliant scientists, and the most powerful politicians are genetically flawed.

Evolutionary Changes in Genomes

The numbers of generations between the present and selected events in human history are shown in Table 5.4. Human and chimp genomes differ in sequence by about 1.5%, or 45 million nucleotides. Assuming

Table 5.4
Numbers of human generations* between the indicated events and the present

Event	Number of Generations
Founding of the United States	8
Time of Mohammed	48
Time of Jesus	68
Time of the Buddha	83
First civilizations	200
Extinction of Neanderthals	1,300
Modern human migration out of Africa	2,000
Origin of anatomically modern humans	6,700
Human-Neanderthal common ancestor	20,000
Human-chimp common ancestor	230,000

*Assumes one generation is 30 years.

230,000 generations in both the human and chimp lines since the common human-chimp ancestor, 1,000 mutations per generation, and an average reproductive population size for both species of 10,000, there are approximately 4.6 trillion possible sequence differences between humans and chimps—nearly 1,000 times the size of the entire human and chimp genomes. Clearly, therefore, only a minuscule fraction of the variants that arise from mutation survive over many generations. Genomes are conservative; they evolve very slowly.

From a long-term genetic perspective, humanity can be considered to be a collection of DNA molecules that have been passed down from one generation to the next. Our bodies last a maximum of about 100 years, while DNA molecules live with only slow, gradual change for many millions of years. Genetically speaking, our bodies serve only to pass DNA molecules on to the next generation (Figure 5.9).

Figure 5.9 Generations

In Chapter 3, I described how the relentless process of mutation increases DNA sequence diversity in human populations. But there are also countering forces, especially including random genetic drift and population bottlenecks, that reduce sequence diversity. Drift tends to eliminate rare variants and fix abundant variants. Drift is usually a slow process, but speeds up markedly during temporary reductions in population size, or what geneticists call population bottlenecks (31–34).

There have probably been many population bottlenecks in the human line since the common human-chimp ancestor. Almost nothing is known about most of these, but geneticists do know that when modern humans successfully migrated out of Africa starting about 60,000 years ago, they experienced a number of bottlenecks (34). We can readily imagine that during this migration, relatively small groups of people moved from their homeland to colonize an area farther along a coast or into the next mountain valley. These small groups carried with them only a fraction of the genetic variation present in the homeland population and therefore constituted bottlenecks.

As a result of these bottlenecks, DNA diversity in all Out-of-Africa populations is lower than in Sub-Saharan African populations (34–36). Many relatively rare sequence variants present in Sub-Saharan Africans are entirely absent from Out-of-Africa populations. In addition, the greater the geographic distance between a human population and Africa, the lower the sequence diversity. Native Americans and Polynesians, for example, have lower sequence diversity than other major human populations (34–35). While mutation is currently increasing sequence diversity in all human populations, it has not yet compensated entirely for the loss of diversity from the migration out of Africa.

Nearly all common human variants are neutral. In Chapter 3 (Table 3.4), I described how neutral variants are free to randomly increase or decrease in populations. As modern humans spread out over the globe, they became reproductively isolated. Neutral variants increased or decreased in frequency independently in each population (see *HBB* nucleotide 9 in Table 5.3). By analyzing variants that differ in frequency among populations, it is possible, *using DNA analysis alone*, to determine the geographic origin of a person's ancestors (what I call geo-ancestry) (37). This is the basis of the DNA ancestry industry, which has become quite large. By the end of 2024, about 30 million people worldwide had their DNA tested for ancestry.

Summary

- Life has existed on our planet for billions of years.

- Very gradually, over long periods of time, initial, simple organisms evolved into more complex organisms, including eventually plants and animals.
- Anatomically modern humans are a young species, a mere 200,000 years old.
- Our closest living relatives are chimpanzees. The common ancestor of humans and chimps lived approximately seven million years ago.
- Modern humans arose in Africa and migrated out of Africa about 60,000 years ago to colonize the rest of the world.
- During this migration, our ancestors encountered Neanderthals living in Europe and Asia.
- Our ancestors were likely a primary cause of Neanderthal extinction, but during our relatively brief overlap, humans and Neanderthals interbred such that all or nearly all humans living today, except those from Sub-Saharan Africa, carry a small amount of Neanderthal DNA in their genomes.
- When modern humans migrated out of Africa, they experienced population bottlenecks. As a result of these bottlenecks, many rare DNA variants that are present in Africans were lost, and sequence diversity was reduced.
- All humans living today, from anywhere around the planet, differ by only about 0.1% in their DNA sequences. These small differences are insufficient to split humans into different species.
- The genomes of all species, including humans, evolve, but only very slowly over many generations.
- From a genetic perspective, humanity is a collection of DNA molecules that have been passed down from one generation to the next.
- There is no such thing as a genetically perfect person. All people have many deleterious variants in their genomes and are therefore genetically flawed.

Chapter 6

APPLICATIONS OF CLINICAL DNA TESTING

Major Areas of Clinical DNA Testing

DNA testing in healthcare (clinical DNA testing) is a relatively new activity. It started slowly in about 1970, but is now expanding rapidly. Clinical DNA testing is performed in three major areas:

- testing for the microorganisms that infect our bodies
- testing of tumor DNA in cancer patients
- testing of germline DNA for heritable DNA variants

Most of this chapter, and nearly all of this book, is focused on the third area, testing of germline DNA. Germline DNA is the DNA that is present in all of our cells and that is passed from one generation to the next. I will, however, briefly describe the other two major areas of clinical DNA testing.

Like all species, the microorganisms that infect our bodies have their own genomes. DNA testing for pathogen genomes has become a powerful research tool, and is also finding wide application in clinical diagnostics. For instance, during the 2020–2022 COVID-19 pandemic, most patients in developed countries were diagnosed only after detection of the viral genome in their bodies. Billions of COVID-19 DNA tests were performed worldwide (1). Sequencing of millions of complete virus genomes also helped epidemiologists track the rise and fall of different virus variants (Alpha, Beta, Delta, Omicron, etc.) (2–3).

Testing for the DNA variants that arise via somatic mutation in tumor cells has two primary applications. The first is screening for early detection of cancer (4–6). This involves DNA testing of stool and, more recently, blood. As most know, cancer is far more treatable when it is caught at an early stage. Screening for early-stage cancer is a marvelous example of preventive medicine.

The second application of tumor DNA testing is for advanced cancer patients (7–9). The goal here is to tailor treatments to the specific somatic DNA variants present in the tumors. Different cancer patients, even those with the same affected organ, have different somatic variants. The optimal treatment for the patient depends upon the nature of these variants. This is a great example of precision medicine (10). Precision medicine means that instead of treating an entire group of patients, like all breast cancer patients, the same way, treatment is specified by the variants that are present in individual patients. Clinical DNA testing is required for most forms of precision medicine.

Primary Applications of Clinical Germline DNA Testing

The primary healthcare applications of germline testing are listed in Table 6.1. Each of these important, overlapping applications is discussed below. As you'll see, *everyone* can benefit from germline DNA testing, regardless whether the person is healthy, severely ill, or anywhere in between.

Table 6.1

Primary applications of germline DNA testing in healthcare

- Diagnosis of disease and disability
- Prevention of disease and disability
- Reproductive planning
- Pharmacogenetics
- Research

Diagnosis of Disease and Disability

Much of germline clinical DNA testing is currently performed for the purpose of diagnosis. An accurate diagnosis is required for optimal treatment and management. The absence of a diagnosis or an incorrect diagnosis often leads to ineffective, wasteful, and sometimes even harmful treatments.

Unfortunately, incorrect diagnoses are quite common in healthcare. A large study on diagnosis from 2015 by the US Institute of Medicine (now National Academy of Medicine) (11) reached the following conclusions:

- Most people will experience at least one diagnostic error in their lifetime, sometimes with devastating consequences.
- 5% of US adults who seek outpatient care each year experience a diagnostic error.
- Diagnostic errors contribute to approximately 10% of patient deaths.
- Diagnostic errors account for 6%–17% of hospital adverse events.
- Getting the right diagnosis is a key aspect of healthcare—it provides an explanation of a patient's health problem and informs subsequent healthcare decisions.
- Improving the diagnostic process is not only possible, but also represents a moral, professional, and public health imperative.

A more recent study reported that about 800,000 deaths or cases of permanent morbidity annually in the US are the result of diagnostic error (12).

While some diagnostic errors are due to human mistakes, most occur because current knowledge and understanding are simply insufficient to determine what is wrong with the patient. In many cases, even after years of costly examinations and testing (a diagnostic odyssey), the patient still remains without a correct diagnosis (see Case 2 in Chapter 3 and also Case 3 below).

There are huge numbers of examples that demonstrate why optimal treatment depends on a correct diagnosis. I'll present just three, all of which also nicely exemplify precision medicine.

Epileptic seizures in an infant or child can have negative long-term health consequences. Healthcare providers therefore attempt to control the seizures as early in life as possible. But there are many drugs that can be prescribed for this purpose. So which drug should be used? Seizures in children may be caused by deleterious variants in many different genes. The best drugs to be prescribed depend upon the gene involved (Table 6.2) (13–21). Of course, identifying the gene involved requires clinical DNA testing. The newborn Alesandro in Case 1 in Chapter 2 had deleterious variants in both of his *ALDH7A1* genes. Pyridoxine (vitamin B6) was therefore the most effective treatment for Alesandro. Note also that for children with deleterious variants in *SCN1A*, some drugs are harmful.

Table 6.2
Optimal drugs for treating children with epilepsy

Gene Involved	Drug(s)
ALDH7A1	pyridoxine (vitamin B6)
CHRNA4	zonisamide (for p.Ser284Leu variant)
KCNQ2	carbamazepine, phenytoin
KCNT1	stiripentol, clonazepam, quinidine
PRRT2	carbamazepine
SCN1A	clobazam, stiripentol, diazepam, clonazepam, valproate **Avoid:** carbamazepine, lamotrigine, and vigabatrin
TSC1, TSC2	vigabatrin

Cystic fibrosis (CF) is one of the most common recessive single-gene disorders in people with European ancestry (Tables 4.3–4.5) (22–23). Roughly 1 in 50 Europeans are carriers for a deleterious CF variant in the *CFTR* gene. A century ago, CF patients rarely lived into their 20s. Today, due to improved treatments and better overall diet and health, CF patients often live into their 50s and beyond. Over 2,000 different deleterious variants in the *CFTR* gene have been identified (a few of these are listed in Table 4.4). A number of drugs have recently been developed that are useful in the treatment of CF (for example ivacaftor, lumacaftor, tezacaftor, elexacaftor), but these drugs are only effective in patients with *specific* deleterious variants. For instance, the combination drug ivacaftor-tezacaftor is recommended in children more than 6 years of age who are homozygous for the most common deleterious variant: c.1521_1523del (p.Phe508del) (22).

The concept of correcting an abnormal gene in patients with single-gene disorders, called gene therapy, has been around now for decades (24). In just the last few years, however, gene therapy has finally moved from research labs to clinics. Approved gene therapies are now available for patients with sickle cell anemia, a form of blindness, spinal muscular atrophy, and hemophilia, and many additional gene therapies for a wide range of single-gene disorders are under active development. Each gene therapy is appropriate only for those patients with abnormalities in one specific gene. As an example, the first approved gene therapy was for a rare, recessive form of blindness due to abnormalities in the *RPE65* gene (25). But because retinal blindness can be caused by deleterious variants in over 300 different genes, clinical DNA testing must be performed to determine whether the *RPE65* gene therapy is warranted.

DNA testing in many cases assists healthcare providers reach quick and accurate diagnoses. Thousands of articles in research journals have now confirmed this fact. References 26–34 are just a tiny, but hopefully reasonably representative, sampling of this vast literature. The conclusion that DNA testing aids diagnosis is also confirmed by the direct experiences of clinical testing laboratories. Millions of patients have now been successfully diagnosed through clinical DNA testing.

Case 3: A Diagnostic Odyssey

John, age 60, had intermittent health problems his entire adult life. Ever since his college days, he had been experiencing bouts of severe abdominal pain and nausea. The first episode occurred when he was at a party with his friends and drank too much. Other episodes occurred after he went on a diet, and when he developed a bad sinus infection. These bouts of illness were costly and disruptive to his life. Twice he was hospitalized. At age 56 he was forced to quit his job. Despite being examined numerous times, John never received a correct diagnosis.

Finally, his provider ordered a genome DNA sequencing test. In two weeks, the lab report came back, showing that John was heterozygous for a deleterious variant in his *HMBS* gene. This variant was the clear cause of his disorder: acute intermittent porphyria (35).

To my knowledge, there is not yet any cure for this dominant form of porphyria, but it has been known for many years that the bouts of illness are triggered by alcohol, some infections, dieting, and a number of drugs. If John had received his DNA diagnosis early in life, he could have avoided these triggers, and his life would have taken a dramatically different course.

Also, as a consequence of his testing result, John's two children and his sister were tested. His sister had experienced similar symptoms during her life. Not surprisingly, John's sister and one of his children were found to be heterozygous for the same deleterious variant. Both were then counseled to avoid the triggers.

Unfortunately, porphyria is far from the only disorder with missed diagnoses in the absence of DNA testing. Many cases of heart disease (36–37), single-gene subforms of cancer (38), and neurological disorders (39) can today be easily diagnosed through DNA testing, yet often go undiagnosed through standard healthcare (40). And this is certainly just a small sampling of a very large problem.

Sick newborns in neonatal intensive care units (NICUs) are a current focus of clinical DNA testing for the purpose of diagnosis. Much recent research has demonstrated the utility of exome and genome sequencing for this group of patients (33–34). Slowly, genome sequencing is becoming standard of care for these infants. But I think this practice needs to be extended to all patients with serious health problems. In my opinion, there is no excuse today for allowing any patient to go undiagnosed or to receive an incorrect diagnosis because a DNA test was not ordered. At a minimum, every healthcare provider should know both the highly penetrant deleterious variants and the significant new variants that arose through germline mutation in each patient.

Prevention of Disease and Disability

Preventing disease and disability is nearly always better than dealing with the problem after it has arisen. The old saying attributed to Benjamin Franklin that "an ounce of prevention is worth a pound of cure" really is true. Making a little effort and paying a little money to prevent disease is vastly preferable to dealing with the enormous costs that accompany disease. And disease encumbers not only the direct costs of treatment, but also other significant costs, such as strains on relationships, sidetracking of careers, and shorter and lower-quality lives. Through several avenues, DNA testing is advancing disease prevention.

I include amelioration of disease as an important component of disease prevention. In many cases, it is not possible today to completely prevent disease, but it is possible to delay the onset or reduce the severity. Amelioration should not be undervalued. To a 76-year-old who is beginning to develop Alzheimer's disease, the difference between losing mental function at age 78 versus 86 is huge.

There are now thousands of examples of the use of DNA testing for disease prevention. In nearly all cases, if a person is known to be at high risk for a disease, then steps can be taken to at least ameliorate the disease (41). This applies to both single-gene and complex disorders. John's acute intermittent porphyria in Case 3 is a nice example of the prevention of a single-gene disorder through DNA testing. Two other examples involve the recessive disease phenylketonuria and single-gene subforms of cancer.

Phenylketonuria (PKU) is caused by deleterious variants in the gene that codes for an enzyme involved in the chemical breakdown of the amino acid phenylalanine (Phe) (42). Much as a dam on a river creates a reservoir, when the breakdown of Phe is blocked, levels of this amino acid in the body increase. High levels of Phe cause intellectual disability. Fortunately, this disability can be prevented by giving the patient a diet that is low in Phe.

Case 4: Cancer Prevention

Jayanti was a vigorous, middle-aged woman and the mother of two teenagers. Although Jayanti was herself healthy, she had a family history of cancer. For her own benefit and that of her children, she underwent testing of her cancer genes to determine whether there was a single-gene subform of cancer running in her family. The lab found that Jayanti had a large heterozygous deletion in one of her *MSH2* genes. This deletion meant that she had Lynch syndrome, a heritable cancer predisposition syndrome involving colon and other types of cancer (43). Her risk of developing colon cancer by age 70 was about 20 times the risk of the average person.

As a result of this DNA diagnosis, Jayanti took a number of preventive actions, including an *annual* colonoscopy and annual scoping of her upper gastrointestinal tract. She also had surgery to remove her ovaries and uterus. With these measures, her chance of developing life-threatening cancer was considerably reduced. Jayanti's children were also tested for the deletion.

Prevention is also starting to be applied to genetically complex disorders. For most genetically complex disorders, risk for the disease in a population can be represented by what statisticians call a normal distribution (Figure 6.1A). Increasing risk for the disease is plotted on the horizontal axis and numbers of people on the vertical axis. The great majority of people have risk that is close to the population average (Figure 6.1B). A relatively few lucky people have especially low risk (Figure 6.1C), and a relatively few unlucky people have especially high risk (Figure 6.1D). If those who are at especially high risk can be identified early in life, before symptoms appear, then available preventive measures may be deployed.

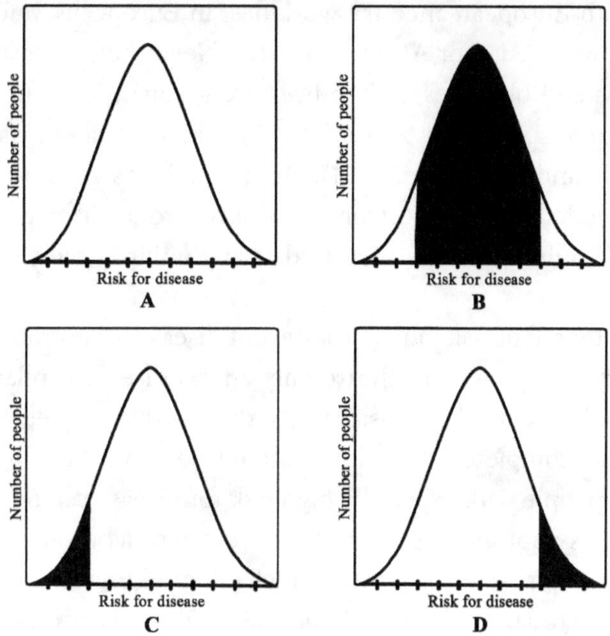

Figure 6.1 Risk distributions for complex disorders and traits
Panel A: Risk distribution. Panel B: The great majority of people with risk near the average. Panel C: People at low risk. Panel D: People at high risk.

In Chapter 4, I described how overall risk for a complex disorder can be estimated by combining many heritable and nonheritable risk factors (Figures 4.8 and 4.9). Algorithms that combine risk factors into a single score are under active development and testing (44–49). Such algorithms yield what are called risk indices. If the algorithms utilize only genetic variants and not environmental risk factors, then they produce what are called polygenic indices (also called polygenic scores or polygenic risk scores). Risk (and polygenic) indices have normal distributions like those shown in Figure 6.1. Many risk indices have now been developed for many different complex disorders and traits.

Risk indices are still in the relatively early stages of development and don't yet do a great job of predicting risk. Not all risk factors have been identified, and researchers don't know the best ways to combine the factors. The indices also work best in the population from which they were derived (44–48). For instance, polygenic indices developed from people with European ancestry work best in Europeans, but less well in people with Asian or African ancestry. Nevertheless, despite these limitations and obstacles, risk indices are gradually moving from research into hospitals and clinics (50–52). Clinical trials of risk indices are already underway (45, 53–54). Once a patient's genome sequence has been added to their electronic health record, a virtually unlimited number of indices can be calculated at any point in their life at very little cost.

The potential of risk indices to prevent disease is enormous. Today, patients typically seek healthcare only after a disease is relatively advanced and causing them distress. By then, although treatment may be possible, complete prevention is often ruled out. With risk indices, however, people with especially high risk for disease can be identified long before symptoms appear, and then prevention becomes a realistic option. Examples of disorders for which effective preventive measures are available today are listed in Table 6.3. And this is just the tip of the iceberg. New and improved preventive measures are continually being developed. I think in future providers will be able to prevent nearly all disorders.

Table 6.3
Examples of disorders at least partially preventable today

Disorder	Preventive Measure(s)
Hemochromatosis	Phlebotomy
Cancer	Enhanced screening; surgery
Coronary artery disease	Statins to reduce cholesterol levels
Type 1 diabetes	Metabolic control; immunointervention
Glaucoma	Topical prostaglandin analogues; laser trabeculoplasty
Macular degeneration	Antioxidants such as lutein; anti-VEGF agents
Alzheimer's disease	Anti-Aβ antibodies

References: hemochromatosis (55); type 1 diabetes (56); glaucoma (57); macular degeneration (58–60); Alzheimer's (61–62).

If a person's risk index for a specific disorder is high, and if relatively safe and inexpensive preventive measures are available, those measures might be implemented solely based on the index. However, I think a more common approach for people with high indices will be to look for the earliest signs of the disease before implementing preventive measures. For instance, because current measures for preventing Alzheimer's disease are expensive and only modestly effective (61–62), these measures might be implemented only for patients who are beginning to demonstrate symptoms of the disease. Fortunately, better screening methods to detect disease at earlier stages are also being developed (see, for example, 63–64).

Reproductive Planning

In addition to prevention through healthcare interventions, thousands of single-gene disorders and aneuploidies can be prevented through reproductive planning. Nearly all couples want healthy children. Caring for a child with a severe illness or disability is very costly in many ways. Reproductive planning through DNA testing involves a number of different approaches, including partner selection, prenatal screening, prenatal testing, and assisted reproductive technologies. Reproductive planning is already reducing the prevalence of a number of disorders, including Tay-Sachs disease, cystic fibrosis, thalassemia, and trisomy 21 (65–68). The topic of reproductive planning is covered in depth in Chapter 11.

Pharmacogenetics

Pharmacogenetics (also often called pharmacogenomics) means to utilize DNA testing to inform drug prescription. In a growing number of cases, based upon the patient's DNA sequences, providers can prescribe the safest, most effective drug at the best initial dose (69–71). Since in developed countries the great majority of people take one or more pharmaceuticals during their lifetimes, pharmacogenetics can benefit nearly everyone. Pharmacogenetics is a key component of precision medicine.

Each year, enormous numbers of patients are prescribed drugs that are ineffective and/or cause adverse reactions. An analysis published in 2015 concluded that for the 10 best-selling drugs in the US at that time, for every one person helped by the drug, 3–24 people failed to improve (72). For many disorders, physicians must choose from multiple drugs that may be prescribed (see, for example, Table 6.2). Without DNA testing, providers often have to try multiple drugs until they find one that is effective.

Adverse drug reactions (ADRs), which can range from mild discomfort to hospitalization to death, are an enormous health problem (71, 73–75). Approximately 15% of all hospital patients experience an ADR. In developed nations, regulatory bodies like the US Food and Drug Administration (FDA) approve drugs only after extensive safety testing. Why, then, do ADRs occur? A good part of the answer is that people differ in their DNA sequences. Over the last few decades, geneticists have begun to learn that a person's genome strongly influences their reactions to drugs (69–71).

DNA sequences influence a person's reaction to drugs through a number of mechanisms. Drugs remain in our bodies for only a limited time. They are usually chemically modified by enzymes and then excreted in the urine and/or feces. Variations in the genes that code for these metabolizing enzymes affect the concentration of the drug in the body over time. Some people have variants that result in unusually slow drug metabolism. Drugs remain in their bodies at higher concentrations for longer periods of time, which can cause an effect similar to an overdose. Others have variants that result in unusually rapid drug metabolism. This can reduce the effectiveness of a drug.

The protein target(s) of a drug may also be altered through mutation such that the drug is ineffective or causes an adverse reaction. Tumor cells in cancer patients often evolve resistance to chemotherapeutic agents through somatic mutation. An example involving a deadly ADR is described in the following case.

Case 5: Deadly Adverse Drug Reaction

Jason was a 17-year-old high school student who loved to play sports. His favorite sport was American-type football. During one of the games, Jason injured his shoulder. His doctors recommended surgery to repair the injury. Jason and his parents arrived at the hospital for what they expected would be a routine, relatively minor surgical procedure.

Unfortunately, when Jason was given standard anesthetic, his body temperature zoomed and all the muscles in his body contracted. Despite intense efforts, his physicians were unable to save him. Jason died as a result of a rare, severe adverse reaction to commonly used anesthetics called malignant hyperthermia (MH).

Jason's parents were, of course, devastated by his death. But later, when they had begun to recover, it was recommended that they both receive DNA testing for MH to learn whether they were susceptible. Many cases of MH are known to be caused by rare variants in the *RYR1* gene, which codes for a key protein in skeletal muscle. These variants do not affect the muscle function of the RYR1 protein and therefore usually go undetected. The DNA testing revealed that Jason's father was heterozygous for the *RYR1* c.7300G>A (p.Gly2434Arg) variant. This variant is known to cause susceptibility to MH and was very likely the cause of Jason's death.

Fortunately for those who are susceptible to MH, alternative anesthetics can be used that do not cause the adverse reaction. As a result of his testing, Jason's father began wearing a medical alert bracelet that warned providers he was susceptible to MH. If Jason had received genome sequencing early in life, then the dangerous MH variant would easily have been identified, and he would likely be alive and healthy today.

Recently, outcomes of clinical pharmacogenetics trials have begun to be published (76–80). These trials have consistently shown that pharmacogenetics improves healthcare. In probably the largest clinical trial completed to date (80), testing of just 12 pharmacogenetics genes reduced ADRs by about 25%. And there is considerable room for further improvement. Much more needs to be learned about the variants that affect our reactions to drugs, and the ways in which these variants interact. My opinion is that the field of pharmacogenetics is just in its infancy. Already, pharmacogenetics is beginning to improve healthcare, and in future, as knowledge grows, it will have a much greater impact.

Research

These days, many of the biggest advances in human genetics are being made by researchers who analyze both the DNA sequences and the clinical features of large numbers of people (10, 81–82). Through such studies, many deleterious DNA variants that cause or increase the risk for disease and disability are being identified and characterized. Also through such studies, polygenic indices are being improved, and new drugs are being developed.

We thus have a virtuous cycle. As the sequences from more people are collected and analyzed, DNA testing becomes more effective. And as DNA testing becomes more effective, more people will demand this healthcare service. We can all help to improve healthcare by sharing our de-identified DNA and health information with researchers. I think we all should.

Other Healthcare Applications of Germline DNA Testing

Germline testing has several useful applications beyond those listed in Table 6.1. Among these are identification of individuals, construction and confirmation of family trees, determination of geoancestry, HLA (human leukocyte antigen) typing for transplantation, blood typing, and detection of sequencing errors.

DNA testing can be used, much like a fingerprint, to identify an individual. Perpetrators of crimes are frequently identified these days by small amounts of their DNA left at crime scenes (83). In the US, a large database of DNA testing results from convicted criminals and some accused individuals is maintained by the Federal Bureau of Investigation (FBI). As of 2024, the FBI CODIS database contained DNA data from about 25 million individuals. DNA testing is also often used to identify human remains. Many of the remains from the 9/11/2001 terrorist attack on the skyscrapers in New York City were identified using DNA testing (84). Identifying such remains is important because it brings closure to the relatives of the deceased.

Because each person inherits half their DNA from their mother and half from their father, the variants identified through genome sequencing can readily be used to construct family trees. Accurate family trees are useful in healthcare. Family history of disease, for example, is one of the strongest risk factors for many disorders (see Case 7 in Chapter 7). The family trees that people assume to be correct are good starting points, but are sometimes inaccurate mainly due to false paternity (the assumed father is not the biological father). If birth years and genome sequences from a child and his/her biological parents are available, then even if this information is buried in a database of many millions of people, geneticists can easily connect the child to the parents.

As described in the previous chapter, it is also possible, using genome sequence information alone, to determine the geoancestry of a person. I define geoancestry as the continental origin of a person's ancestors as of about the year 1500. In addition to being of personal interest, geoancestry is also useful in healthcare. Polygenic indices can be tailored, for instance, to a person's geoancestry.

These days, most HLA typing, which is essential for transplantation, is accomplished through DNA testing (85). A special DNA test just for this purpose is usually performed. But the same information can also be extracted from a genome sequence. If genome sequencing is performed early in life, and if the data are retained (see the "Data" section in the next chapter), then the HLA results can easily be recalled if and when they are needed later in life. Similarly, blood types can also be determined using DNA testing (86). Knowing blood types is important for transfusions and also for maternal-infant Rh factor compatibility.

Finally, sequencing errors in accredited clinical labs, while rare, do occasionally occur. Using family structure and sequence data from family members, geneticists can often detect sequencing errors. A simple example is shown in Figure 6.2. Because the paternal grandfather and the grandson both carry the *rare* G variant, it is very likely the father also carries this variant, and the apparent sequence for the father from the laboratory is incorrect.

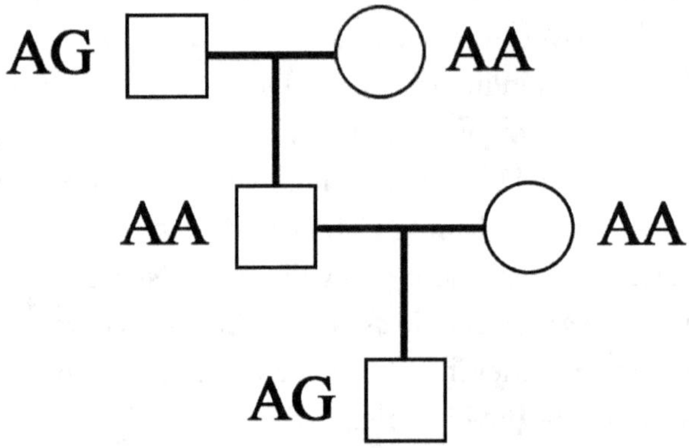

Figure 6.2 Detection of sequence errors through DNA testing of family members
The "G" variant is rare. Because both the paternal grandfather and the grandson carry the G variant, it is likely that the DNA test for the father is incorrect, and that he is actually heterozygous AG.

Summary

- Because of its great benefit to patients, clinical DNA testing is expanding rapidly.
- There are three main areas of clinical DNA testing:
  - Testing for the microorganisms that infect our bodies
  - Testing of tumor DNA in cancer patients
  - Testing of germline DNA
- The primary applications of germline DNA testing are:
  - Diagnosis

  - Disease prevention
  - Reproductive planning
  - Pharmacogenetics
  - Research
- DNA testing is a powerful tool to help healthcare providers reach rapid and accurate diagnoses. An accurate diagnosis is essential for optimal healthcare.
- Through DNA testing and risk indices, geneticists' capability to predict a person's risk for developing a specific disorder is improving rapidly. Gauging a person's risk for disease is a key step in prevention.
- A significant fraction of disease and disability, particularly single-gene disorders, can be avoided through reproductive planning (Chapter 11).
- Geneticists are still learning a great deal about pharmacogenetics, but even so, pharmacogenetics is already helping to reduce adverse drug reactions and improve the efficacy of drugs.
- The more DNA sequence and clinical information that becomes available to researchers, the more rapidly researchers will learn, and the more beneficial clinical DNA testing will become.
- Additional important applications of clinical DNA testing include individual identification, creation of family trees, and determination of geoancestry.
- Clinical DNA testing benefits everyone.

Chapter 7

CLINICAL DNA TESTING FACTORS

Limitations of Clinical DNA Testing

In the previous chapter, I presented the many beneficial applications of clinical DNA testing. But clinical DNA testing is not a panacea. For the sake of impartiality, I think it's important to describe what I see as the current major limitations.

In patients with a significant health problem or disability, even full genome sequencing yields a clear diagnosis only about a third of the time (1–6). Partly this is because of geneticists' limited ability to interpret the sequences. Due to lack of evidence, many rare sequence variants currently have unknown clinical significance. Also, geneticists' ability to predict the combined effect of two or more rare variants in different genes is almost nil. Failure to reach a diagnosis is sometimes due to technical limitation. There are important regions of the genome that are today difficult to sequence. Environmental causes of disease are another important reason why diagnosis is not always achieved. Diagnostic rates through DNA testing will improve in future but, especially because of environmental factors, will never reach 100%.

Geneticists today do not have good estimates of penetrance for most deleterious variants. This can lead to rare mistakes. For instance, a woman may undergo a mastectomy solely because she is heterozygous for a breast cancer risk variant with 5% penetrance. Similarly, healthy fetuses may be aborted solely because they carry a low-penetrance risk variant.

As described in the "Data" section below, it's important for healthcare purposes to retain patient DNA sequence information. Yet patient sequences are just beginning to be entered into patient electronic health records, and are often not retained today.

Too few healthcare providers have special training and expertise in genetics. As shown in Table 7.1, there are three major groups of healthcare personnel with expertise in clinical genetics.

Table 7.1
Numbers of healthcare providers with genetics expertise in the US

Group	Current Numbers Practicing in US	Number Added Each Year by Training Programs
MD medical geneticists	700	100
Genetic counselors	7,500	600
PhD lab geneticists	1,200	130

Physicians with special training in genetics—usually called medical or clinical geneticists—order DNA tests and diagnose and manage patients. They also often consult with other providers, lead testing laboratories, and are involved in research. The number of practicing medical geneticists in the US (about 700) (7–8) is very small compared to, for example, pediatricians (33,000) or general practitioners (168,000) (9). Some less populous US states today have no practicing medical geneticists.

Genetic counselors specialize in communicating genetic information to patients and providers. Here, numbers are more encouraging (10–12). Training programs for genetic counselors have grown both in number and class size in recent years. In the US alone, about 600 new genetic counselors are graduating from training programs each year.

Laboratory clinical geneticists (usually PhDs) supervise clinical DNA testing and interpret test results. Numbers of these specialists appear to be growing in the US, but not as fast as needed (13).

The number of healthcare genetics specialists worldwide is also growing, but only slowly (7, 12). Numbers of these specialists per capita in other nations are even lower than in the US (12–14). Some developing nations have no such specialists at all, and even in some populous countries like India, the numbers are currently tiny.

Because there are too few genetic specialists to serve entire populations, many healthcare providers who are not genetic specialists will need to deal with genetics. Today, however, many providers avoid doing so because they don't understand genetics well and feel it's too complex. Patients also often shy away from genetics for the same reasons. Genetics education therefore needs to be improved for providers and patients alike. This book is one small part of that effort.

Ethical, Legal, and Social Implications of Germline Testing

In 1990, near the start of the Human Genome Project (see below), the US National Institutes of Health (NIH) set aside funds to study the ethical, legal, and social implications (ELSI) of the Genome Project and human genetics more generally (15). Many ELSI studies were subsequently completed and published. These studies identified a number of potential risks associated with germline testing, some of which are listed in Table 7.2.

Table 7.2
Some of the major ELSI concerns associated with DNA testing

- Privacy of genetic testing results
- Discrimination in employment
- Discrimination in health and other insurance
- Stigmatization of groups such as minorities and the disabled
- Psychological distress

Fortunately, the dire warnings of ethicists have not materialized. Tens of millions of people worldwide have now had their DNA tested without triggering the hypothetical problems listed in Table 7.2. People who have had DNA testing are not more emotionally distressed than those who have not had testing (16–18). The US and other countries have passed laws prohibiting genetic discrimination in employment and insurance (19). Examples, especially recent examples, of individuals who have been harmed through DNA testing are vanishingly rare. My own opinion is that for most people, having your genome sequenced is less frightening than receiving a mammogram or colonoscopy. Genome sequencing might tell you that you have a *predisposition* for cancer, but a mammogram or colonoscopy might tell you that you *have* cancer.

This does not mean, however, that we should be cavalier about DNA testing. DNA sequence information is potent. It informs risk of disease, reproductive outcomes, and ancestry. Many thorny questions are raised by DNA testing, a small sampling of which are listed in Table 7.3.

Table 7.3
Some difficult ethical questions raised by DNA testing and human genetics

- Should adopted individuals be provided with the identities of their biological parents?
- Should people be permitted to withhold their own DNA sequence information from being used anonymously to benefit their relatives?
- Is it a violation of privacy to sequence DNA from ancient human remains, especially those that have been identified, such as remains of famous people?
- Should DNA sequence information in *clinical* DNA databases be used to solve crimes?
- Should sperm donors be permitted to father unlimited numbers of children?
- Should germline DNA editing be permitted?
- Should people be cloned?

I believe DNA sequence information should be treated with great respect and kept strictly confidential; DNA sequences should be among the most private of all personal information. But I also think that the benefits of DNA testing are so great that the inherent controversies must be overcome. Answering difficult ethical questions such as those listed in Table 7.3 will require lengthy and careful deliberation, but such questions can, and will, be answered. Resolving difficult ethical questions is nothing new. Human societies have been doing it for thousands of years.

What Type of Germline Testing?

Over the last few decades, a number of laboratory methods have been developed for analyzing the human genome. Several of these methods are still in use in clinical labs, but the current favored method is DNA sequencing. Sequencing of the first human genome was completed in 2001 as the primary goal of the Human Genome Project (20). This first genome sequence was expensive, costing in the neighborhood of $3 billion. The high cost was largely due to the use of older sequencing technology. Since then, DNA sequencing technology has improved radically, and costs have dropped over a million-fold. A person's complete genome can now be sequenced for about $200, and sequencing technology continues to improve.

Healthcare providers can order many different types of germline sequencing tests for their patients. They can order a test of a single nucleotide to determine whether a patient carries a specific deleterious variant (usually one that has already been found in another family member). They can also order sequencing of a gene panel. If their patient is deaf, for example, they may order sequencing of a panel of about 200 genes that are known to be involved in hearing loss. They can also order sequencing of the exons of all 20,000 genes (an exome sequencing test). Or, they can order sequencing of the entire genome (a genome sequencing test). So, what type of germline test is best?

To help answer this question, I'll recount an old Indian fable.

> A group of blind men lived in a rural village in India. To pass the time, the men often talked and argued among themselves. One of their favorite topics was the fabulous animal, the elephant. There were no elephants in their village.
>
> One day, as a special treat, the blind men were led to the raja's palace grounds where they encountered an elephant. One of the blind men touched the elephant's ear and said: "An elephant is like a great fan." Another touched the elephant's trunk and said: "No, you're wrong, an elephant is like a big snake." A third touched the elephant's tusk and said: "The elephant is like a giant spear." And so they argued all the way home and long thereafter.
>
> The moral of the story, of course, is that people need to experience the whole object for understanding. Just experiencing a portion of the object is often misleading.

So it is with germline DNA sequencing. When geneticists sequence just a single gene or even hundreds of genes, they often get a distorted view of the patient's situation. Even the best medical geneticists cannot always predict which genes are involved in the patient's problem. And because expert medical geneticists are scarce, more and more providers who are not genetic experts are ordering DNA tests. Full genome sequencing for patients bypasses the need to make guesses about which genes might be involved. Genome sequencing covers all people and all disorders.

Genome sequencing also often reveals totally unanticipated health problems. Consider the following case.

> **Case 6: Unanticipated Diagnosis**
>
> Zahra, a four-month-old girl, had congenital heart disease. In an attempt to reach an accurate diagnosis, her provider ordered genome sequencing. The sequencing revealed that Zahra was heterozygous for a large three-million-nucleotide deletion on chromosome 22 (21). This deletion was very likely the primary cause of her heart disease.

> But the sequencing also revealed that Zahra was heterozygous for deleterious variants in both of her *ATP7B* genes on chromosome 13. These variants meant that she had a second, unrelated disorder called Wilson disease (22). In this disease, levels of copper build up to toxic levels in the body and cause irreversible damage to a number of organs, particularly the brain and liver. Fortunately, there is a treatment called chelation that removes the copper from the body. If Zahra can overcome her heart disease, it will be important for her providers to monitor her copper levels and perform the chelation or other therapy as needed.
>
> If Zahra's provider had ordered sequencing of just known heart disease genes, then the Wilson disease diagnosis would have been missed. Only because all the genes were sequenced was it possible to make this diagnosis.

Multiple genetic diagnoses, such as in Zahra's case, are relatively common, occurring in roughly 5% of patients with significant health problems (23–25). In addition, many people are totally unaware that they are at high risk for developing a preventable disorder like cancer later in life. Genome sequencing uncovers such risks.

Many limited, specialized germline DNA tests are carried out today for purposes other than diagnosis. Large numbers of DNA tests are performed *solely* for pharmacogenetics, reproductive planning, cancer predisposition, geoancestry, HLA typing, and other purposes. Genome sequencing serves all of these applications and much more. It is a universal test for all parts of the genome.

Another common reason today for performing limited germline tests is that healthcare funders, such as health insurance companies in the US or provincial governments in Canada, assume they will save money by paying only for tests that cover a portion of the genome. In my experience, this is false economy. The difference in cost between gene panel sequencing and genome sequencing is dropping. Also, if the gene panel test is negative, it is often followed by genome (or exome) sequencing. This ultimately increases total patient costs and prolongs diagnostic odysseys (see Cases 2 and 3) (26–28). In patients who are severely ill, like infants in NICUs, *rapid* diagnosis is especially important (29). Such babies need immediate help; they cannot wait for a diagnostic odyssey.

A substantial, but gradually shrinking, group of geneticists still today prefers exome to genome sequencing. Remember that exome sequencing involves sequencing only about 1.5% of the total nucleotides in the genome. This makes exome sequencing less expensive than genome sequencing. However, genome sequencing has substantial advantages over exome sequencing, as shown in Table 7.4 (30–34). These advantages are a bit technical, but to summarize, genome sequencing provides information on sequence variants located outside of exons that are important in diagnosis and for generating polygenic indices. Genome sequencing is also superior in detecting large deletions and insertions and analyzing some of the more difficult regions of the genome.

Table 7.4
Advantages of genome sequencing over exome sequencing

- Detection of highly penetrant deleterious variants for single-gene disorders that are located outside of exons
- Superior detection of large deletions and insertions of DNA
- Superior analysis of tandem repeats
- Detection of low-penetrance noncoding risk variants
- Superior analysis of genes present in the genome in multiple copies

Michael Astion, a physician leader in laboratory medicine at Seattle Children's Hospital, introduced me to the following equation:

$$\text{Value of a Clinical Test} = \frac{\text{Quality of Test}}{\text{Cost of Test}}$$

The value to patients of a particular laboratory test is the quality of the test divided by its cost. Value is improved either by increasing the quality of the test and/or by decreasing the cost. Based on this equation, and considering the information provided above, I strongly feel that genome sequencing provides the best value to patients. No other germline test provides even close to the amount of information for the price. Therefore, to return to the question of what type of germline test, I conclude that the answer is full *genome sequencing*.

Who Should Receive Genome Sequencing?

Another important question in clinical DNA testing is who should receive genome sequencing. Should it be only those with significant disease or disability? Should it be all newborns? Or should it perhaps be everyone?

Cost is an important factor in answering such questions. Today, a human genome can be sequenced for about $200. However, when the costs of drawing and shipping blood, extracting DNA, interpretating the sequence, and generating test reports are added, the total cost for clinical genome sequencing is in the range of $1,000. That is certainly a good amount of money, but compared to many other common expenditures such as a new cell phone ($900), a new car ($40,000), or raising a child ($250,000), genome sequencing is not especially pricey. In the US, the current average lifetime cost of healthcare is approaching $1 million (35). Genome sequencing, even after adding in the cost of counseling and repeated interpretation of the sequence (see below), is less than 0.3% of this total. Other developed nations spend less per capita on healthcare than Americans, but the costs for genome sequencing would still be less than 1% of lifetime healthcare costs. In my opinion, this is a pittance compared to the vast benefits received.

Healthcare economists have repeatedly concluded that for people with significant healthcare problems, especially sick infants and children, genome sequencing is cost effective (36-40). Fewer studies have been reported on the cost effectiveness of germline testing in healthy individuals, but even here the results have been encouraging. Germline testing has been reported to be cost effective for prevention of breast/colon cancer and hypercholesterolemia (41-42), skin cancer (43), and recessive disease through carrier testing (44). Given these results, and given that these studies considered pretty much all of the costs of testing, but only a small fraction of the benefits (usually just diagnosis), I personally am certain that population-level genome sequencing will be cost effective. Cost, in my opinion, should never prevent anyone in a developed nation from receiving genome sequencing. It is one of the best bargains people will ever see in their lives.

Healthy individuals can benefit from genome sequencing through reproductive planning, pharmacogenetics, and prevention of disease. Even for those rare individuals who never take any prescription drugs and are never sick, sequencing of their genomes will benefit family members directly (see Case 7 below) and all others indirectly through research. Therefore, my answer to the question who should receive genome sequencing is *everyone*.

When in Life Should Genome Sequencing Be Performed?

Except for somatic mutation, our genomes remain unchanged from the moment of fertilization until the day we die. The genome sequencing performed on a newborn is useful to that same person at age 95. As long as the sequence information is retained, clinical genome sequencing needs to be performed only once in a lifetime.

For over 50 years now, nearly every baby born in the US and other developed nations has received testing within a few days of birth for a panel of single-gene disorders (45–48). The purpose of these newborn screening programs is to catch a limited number of rare disorders as early in life as possible so that treatment can begin immediately.

The single-gene disorder that initiated newborn screening is phenylketonuria (PKU) (49). In PKU, unusually high levels of the amino acid Phe impede normal brain development and lead to intellectual disability. This disability can be prevented if the patient is given a diet low in Phe as early in life as possible. If the disease is diagnosed later in life, it is too late. The intellectual disability cannot be reversed.

Newborn screening panels in the US currently test for up to about 50 different single-gene disorders (46), but screening is not performed today for many other disorders for which early diagnosis is important. Consider autism and autism-like disorders (collectively termed autism spectrum disorders or ASDs). ASDs affect up to 2% of children (50–52). The average age at ASD diagnosis today is four years (53). Early diagnosis is important because interventions are most effective when started early (54–55). Because ASDs are highly heritable, they can often be diagnosed through DNA sequencing (56–58). For ASD patients, therefore, the earlier in life genome sequencing is performed and diagnosis established, the better.

Many geneticists, myself included, advocate complementing and eventually replacing the current newborn screening programs with genome sequencing. Research projects involving genome sequencing for hundreds of thousands of newborns, healthy and ill, are now underway (47). I applaud the direction this research is headed. My answer, therefore, to the question when should genome sequencing be performed is *at birth.*

DNA Banking

DNA banking is the long-term, secure storage of a person's DNA, even after death, for the purpose of future testing. Consider the following case.

> **Case 7: Importance of DNA Banking**
>
> Helen was a 42-year-old mother of two daughters. As shown in her family tree in Figure 7.1, Helen had a family history of cancer. Her mother, aunt, and great-aunt all had breast or ovarian cancer. She was concerned about her own risk for cancer, as well as cancer risk for her daughters. Helen therefore decided to have her genome sequenced.
>
> The lab determined that Helen carried a rare sequence variant in one of the copies of her breast cancer 2 (*BRCA2*) gene. This variant altered one of the amino acids in the coded protein. It is known that deleterious variants in *BRCA2* significantly increase a woman's risk of breast and other cancers, but there was not enough evidence in the biomedical literature or databases to determine whether Helen's specific variant was deleterious or not. The lab

therefore interpreted this variant as "uncertain clinical significance"—not very helpful to Helen or her family.

But wait. What if more members of Helen's family were explored? If the affected women in her family were also found to carry Helen's rare variant, then the probability that it is deleterious would be greatly increased.

Notice in Figure 7.1 that the symbols for Helen's aunt and great-aunt have X's through them. An X means that the person is deceased. If the aunt and great-aunt had banked their DNA, then the DNA could easily be retrieved and tested, but if they did not, Helen would be out of luck.

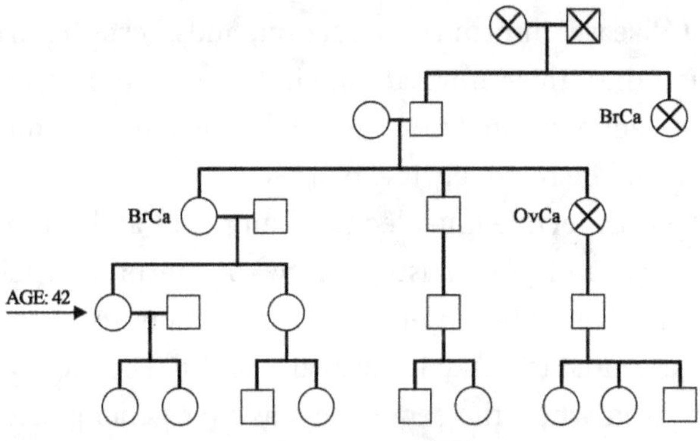

Figure 7.1 Case 7 family tree

The arrow indicates Helen. BrCa is breast cancer; OvCa is ovarian cancer. X's through symbols indicate that the individual is deceased.

Diagnosis of disease and interpretation of sequence variants are both significantly improved by using sequence and clinical information from family members. This is why DNA banking is so important. As mentioned in Chapter 5, geneticists now have the capability to sequence DNA from human remains that are many thousands of years old. Storing DNA in freezers for decades or even a few centuries before sequencing is therefore trivial. DNA banking is also easy and inexpensive. In scale, it can be completed for a *onetime* cost of less than $100.

The importance of DNA banking is not widely recognized. Many healthcare providers today know nothing about it. Nevertheless, DNA banking can be enormously important to families. It's a little like preparing a will—you don't do it for yourself, but rather for your family. After a person has been buried or cremated, it is difficult or impossible to obtain their DNA, so it's best to do the banking while they are still alive. My strong opinion is that everyone should bank their DNA.

Data

To be of the greatest value, the results of germline DNA tests must be saved indefinitely in electronic medical records. There are at least four good reasons for this. First, except for somatic mutation, our DNA sequences do not change over our lifetimes. If stored and retained, the sequence information can be used over a person's entire life to diagnose and prevent disease, inform reproduction, and prescribe drugs. Even after a person dies, this information will benefit family members and others. Most sequence analyses will be completed by computer programs and so will be relatively inexpensive.

Second, our collective knowledge about genes and variants is constantly growing, and geneticists' ability to interpret a person's genome sequence is steadily improving. Periodically reinterpreting the sequence is essential to make it optimally useful. Diagnoses that were originally missed when the sequence was first generated can sometimes be made upon reinterpretation (59–61). Risk indices and computer processing of the raw sequence data are also steadily improving. Portions of the genome that, for technical reasons, are currently difficult to analyze will become tractable in the future.

Third, a person's genome sequence helps family members. Because each child inherits their DNA from their parents, family members are inextricably linked genetically. Even distant cousins share appreciable amounts of DNA. In most of healthcare, the unit of care, as I like to call it, is the individual. When a person comes to a hospital with a broken leg or a severe infection, the unit of care is, as it should be, that individual. But in genetics, the unit of care is not the individual, but the *family* (see all nine cases in this book). To maximize the benefit of genetics for individual patients, providers must have access to sequence information from relatives.

The phase of variants—defined as whether two or more variants within or near a gene are located together on one chromosome or are located on different chromosomes—is often important in healthcare. If, for example, two rare variants in a recessive gene are found in a child, then it is critical to determine the phase of the variants. If both variants are located on the same chromosome, then it is unlikely that they are the cause of disease, but if the variants are located on different chromosomes, then they are much more likely to be the primary cause. In many cases, phase cannot currently be determined simply by sequencing the patient alone. The easiest way to determine phase is to sequence DNA from the parents (Figure 7.2). Phase of variants is also particularly important in pharmacogenetics.

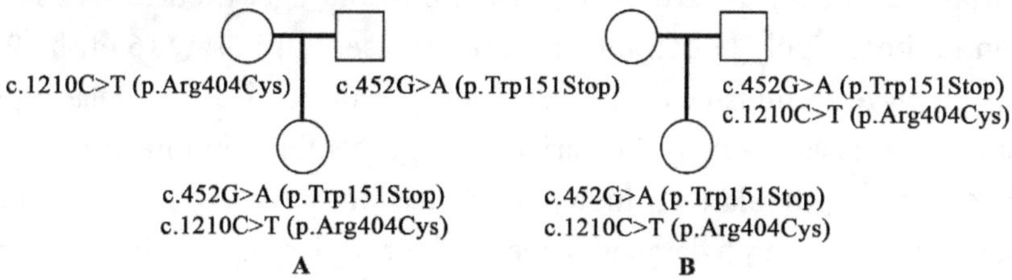

Figure 7.2 Determination of phase using sequence information from parents

The girl in the family is heterozygous for two deleterious variants in the *DHCR7* gene c..452G>A (p.Trp151Stop) and c.1210C>T (p.Arg404Cys). If, as in panel A, the child inherited one variant from each parent, then the variants are on different chromosomes, and the child has a genetic diagnosis of the recessive disease Smith-Lemli-Opitz syndrome. However, if, as in panel B, the child inherited both variants from a single parent, then because the child has one normal chromosome, it is unlikely that she is affected with this recessive disease.

The fourth reason is that sequence information from individuals coupled with their clinical information is vital to advancing human genetics research (62–65). Geneticists use such data to improve knowledge of heritability of disorders and traits, penetrance of variants, clinical significance of variants, polygenic indices, and many other parameters and phenomena. The more data that are available to researchers, the more rapidly clinical genetics and healthcare in general will improve. Data retention is an essential component of the virtuous cycle described in the previous chapter. More data improves the value of genome sequencing, which encourages more people to have their genomes sequenced, which allows researchers to gain new knowledge, and so on, indefinitely.

Many sequence variants are extremely rare, present in less than one per million people. Nearly all highly penetrant deleterious variants are rare, but not all rare variants are deleterious. The trick is to determine which of the many rare variants are a primary cause of disease. To obtain reliable information about such variants, researchers need clinical information from more than just one or two people with the variant. If geneticists can identify, say, 20 people with the same rare dominant variant, and they all live healthy lives into old age, then this variant cannot be a highly penetrant cause of disease. But if, say, 16 of the 20 people have an unusual health problem—for instance a particular type of kidney disease—then the variant can be confidently interpreted as deleterious with relatively high penetrance. Therefore, to provide optimal benefit, research databases need to be very large, containing information from millions, and someday billions, of people.

Especially over about the last century, medicine has made remarkable progress. Many patients who even a few decades ago would have struggled to survive now lead healthy lives. However, one area where our healthcare systems are still deficient is in the reliable, long-term retention of healthcare data. To my knowledge, no nation is yet routinely storing clinical DNA sequences in electronic health records, and no nation is permanently retaining these records.

Should People Wait for Better DNA Testing Technology to Be Developed?

The technology for DNA testing is steadily improving (66). Genome sequencing technology and geneticists' ability to interpret the sequences will have improved appreciably even five years from now. So, should someone who wants to have their genome sequenced wait?

I think the answer to this question is no. Although sequencing technology will undoubtedly improve, it is still pretty good today. And if the raw sequence data is retained, then the sequence can be repeatedly reanalyzed in the future. Once the sequence is obtained, people can begin to benefit from it; waiting postpones these benefits. In addition, none of us knows how long we will live. After a person dies, even if they have banked their DNA, healthcare funders are reluctant to pay for DNA testing. A much better approach is to complete the sequencing while the person is still alive.

Summary

In contrast to preceding chapters, most of the summary points for this chapter, although they are strongly supported by evidence, are still my personal opinions. These opinions are shared by many, but certainly not all, geneticists.

- Genome sequencing will significantly improve healthcare, but there are limitations. Chief among these are the inability to achieve a genetic diagnosis in the majority of cases and a lack of genetics knowledge and understanding among healthcare providers and patients.
- DNA testing does not harm the individuals tested. It does, however, raise some difficult ethical, legal, and social questions.
- Full genome sequencing is the best germline test to perform in most situations. It provides the best value for the patient's money.
- Genome sequencing should ideally be performed at birth.
- Genome sequencing is for everyone because it benefits everyone. The cost is modest, and should not be an obstacle to taking advantage of this valuable service.
- DNA banking is not widely appreciated, yet it is simple, inexpensive, and valuable.
- Healthcare institutions need to do a much better job of retaining patient clinical and DNA sequence data. These sequence data should be retained permanently, and should be available both to providers and anonymously to researchers.
- There are good reasons for individuals to have their genomes sequenced now, as opposed to waiting for improved technology.

Chapter 8

SUMMARY OF PART I, GENETICS AND HEALTHCARE

A Brief Review of Chapters 1–7

In Part I of this book, I described how our DNA molecules serve as blueprints for our bodies. I explained how the information contained in the sequence of nucleotides along the DNA molecules is used to manufacture protein. I portrayed how DNA molecules are transmitted from one generation to the next, and emphasized that mutation is not a rare, freakish event, but rather is a continuous, natural process that occurs between each human generation. I presented a little about the fascinating story of human evolution, and made the case for genome sequencing as an integral and routine component of healthcare.

The summaries at the ends of each chapter provide a quick means of reviewing the major concepts of Part I. I encourage readers to reread these points. It doesn't take long and reinforces learning. To facilitate this step, I have repeated the summary points here. Statements that are my personal opinions as opposed to scientific facts are marked by asterisks (***).

Chapter 2: Human Genetics Basics

- Human genetics is the science of the causes and the inheritance of human variation.

- Proteins are the primary functional components of our bodies. We have about 20,000 different proteins, and enormous numbers of copies of each protein type.
- The instructions for the manufacture of our proteins are contained within our DNA molecules.
- These instructions take the form of the *sequence* of adjacent nucleotides along the DNA polymer.
- The genetic code is simple. Three adjacent nucleotides in DNA code for each amino acid in protein. RNA serves as an intermediate in the transfer of sequence information from DNA to protein.
- A gene is a segment of a DNA molecule that codes for a specific protein. Humans have about 20,000 genes, one for each protein.
- Genome is the total complement of DNA within an individual or species.
- The human genome consists of 46 nuclear DNA molecules and one small mitochondrial DNA molecule.
- In our cells, the long, thin DNA molecules are combined together with proteins. These DNA-protein complexes are called chromosomes. The nuclei of our somatic cells have two copies of each of the 23 different types of chromosomes, 46 chromosomes in all.
- The 23 pairs of chromosomes are split between 22 pairs of numbered chromosomes (the autosomes), and the two sex chromosomes X and Y. Females have two X chromosomes and no Y chromosome, while males have one X and one Y.
- Every time a somatic cell divides, each daughter cell receives the full complement of chromosomes and the full set of DNA instructions.
- Egg and sperm cells (the germ cells) contain only one copy of each chromosome type and half the normal amount of DNA.
- When sperm and egg cells fuse to begin the next generation, the normal number of chromosomes is restored.
- Each person receives one of each chromosome type from their father and one from their mother.
- We have two copies of each gene, except for the genes located on the sex chromosomes in males.

- Each of our approximately 20,000 genes has a specific location along one of the chromosomal DNA molecules.
- Nearly all human genes have multiple coding segments called exons separated along the DNA by introns. Introns are removed from the initial RNA transcripts through a process called splicing.
- Exons comprise only a small fraction, about 1.5%, of our genomes.

Chapter 3: Mutation

- Mutation is defined as a change in the nucleotide sequence of a DNA molecule.
- Mutation is a normal and ubiquitous biological process that occurs in all of our cells throughout our lives and between every human generation.
- Approximately 1,000 mutations occur in each human child. We are all therefore mutants.
- Mutation alters nucleotide sequences in many ways, although some types of change, particularly changes in numbers of tandem repeats and nucleotide substitutions, are much more common than other types of change.
- Roughly speaking, mutations occur at random locations within a genome, and therefore the vast majority occur outside of genes.
- New DNA sequences that arise through mutation are called variants. Nearly all new variants have little or no effect on our health and abilities, but among those with larger effects, deleterious variants greatly outnumber advantageous variants.
- Most new variants vanish from populations within a few generations at most.
- Neutral variants are free to randomly drift up and down in population frequency by chance.
- Natural selection decreases the population frequencies of deleterious variants and increases the population frequencies of advantageous variants.
- Artificial selection is used by human breeders to rapidly change traits in commercial species.
- Mutation has an enormous impact on our species and all other living things.

- Just as we can't prevent hurricanes and earthquakes, we have no ability to stop mutation. Rather, we need to deal with the reality of mutation as best we can.

Chapter 4: Inheritance of Chromosomes, Variants and Traits

- Parents randomly transmit one of their two genes to each of their children. This process is independent for each child.
- Independent segregation of genes and meiotic recombination increase the diversity of genomes.
- Geneticists measure the degree of inheritance of a disorder or trait using a parameter called heritability. Heritability is often estimated through twin or adoption studies.
- Many common health problems and other human traits are strongly influenced by the DNA we inherit from our parents, but are also strongly influenced by nonheritable factors such as environmental exposures and mutation, and hence have heritabilities greater than 0%, but less than 100%.
- Genetically complex disorders are influenced by variants in many different genes and by nonheritable factors.
- Single-gene disorders are predominantly caused by one or two rare, highly penetrant deleterious variants in a single gene. Unrelated individuals affected with the same single-gene disorder usually have different deleterious variants in that gene.
- Single-gene disorders are primarily inherited in simple dominant, recessive, and X-linked patterns.
- Effects of individual risk factors, both heritable and nonheritable, can be combined to estimate a person's overall risk for disease.

Chapter 5: Human Evolutionary History and Genetic Diversity

- Life has existed on our planet for billions of years.
- Very gradually, over long periods of time, initial, simple organisms evolved into more complex organisms, including eventually plants and animals.
- Anatomically modern humans are a young species, a mere 200,000 years old.

- Our closest living relatives are chimpanzees. The common ancestor of humans and chimps lived approximately seven million years ago.
- Modern humans arose in Africa and migrated out of Africa about 60,000 years ago to colonize the rest of the world.
- During this migration, our ancestors encountered Neanderthals living in Europe and Asia.
- Our ancestors were likely a primary cause of Neanderthal extinction, but during our relatively brief overlap, humans and Neanderthals interbred such that all or nearly all humans living today, except those from Sub-Saharan Africa, carry a small amount of Neanderthal DNA in their genomes.
- When modern humans migrated out of Africa, they experienced population bottlenecks. As a result of these bottlenecks, many rare DNA variants that are present in Africans were lost, and sequence diversity was reduced.
- All humans living today, from anywhere around the planet, differ by only about 0.1% in their DNA sequences. These small differences are insufficient to split humans into different species.
- The genomes of all species, including humans, evolve, but only very slowly over many generations.
- From a genetic perspective, humanity is a collection of DNA molecules that have been passed down from one generation to the next.
- There is no such thing as a genetically perfect person. All people have many deleterious variants in their genomes and are therefore genetically flawed.

Chapter 6: Clinical DNA Testing Applications

- Because of its great benefit to patients, clinical DNA testing is expanding rapidly.
- There are three main areas of clinical DNA testing:
    - Testing for the microorganisms that infect our bodies
    - Testing of tumor DNA in cancer patients
    - Testing of germline DNA

- The primary applications of germline DNA testing are:
    - Diagnosis
    - Disease prevention
    - Reproductive planning
    - Pharmacogenetics
    - Research
- DNA testing is a powerful tool to help healthcare providers reach rapid and accurate diagnoses. An accurate diagnosis is essential for optimal healthcare.
- Through DNA testing and risk indices, geneticists' capability to predict a person's risk for developing a specific disorder is improving rapidly. Gauging a person's risk for disease is a key step in prevention.
- A significant fraction of disease and disability, particularly single-gene disorders, can be avoided through reproductive planning (Chapter 11).
- Geneticists are still learning a great deal about pharmacogenetics, but even so, pharmacogenetics is already helping to reduce adverse drug reactions and improve the efficacy of drugs.
- The more DNA sequence and clinical information that becomes available to researchers, the more rapidly researchers will learn, and the more beneficial clinical DNA testing will become.
- Additional important applications of clinical DNA testing include individual identification, creation of family trees, and determination of geoancestry.
- Clinical DNA testing benefits everyone.***

Chapter 7: Clinical DNA Testing Factors

- Genome sequencing will significantly improve healthcare, but there are limitations. Chief among these are the inability to achieve a genetic diagnosis in the majority of cases and a lack of genetics knowledge and understanding among healthcare providers and patients.
- DNA testing does not harm the individuals tested. It does, however, raise some difficult ethical, legal, and social questions.***

- Full genome sequencing is the best germline test to perform in most situations. It provides the best value for the patient's money.***
- Genome sequencing should ideally be performed at birth.***
- Genome sequencing is for everyone because it benefits everyone. The cost is modest, and should not be an obstacle to taking advantage of this valuable service.***
- DNA banking is not widely appreciated, yet it is simple, inexpensive, and valuable.***
- Healthcare institutions need to do a much better job of retaining patient clinical and DNA sequence data. These sequence data should be retained permanently, and should be available both to providers and anonymously to researchers.***
- There are good reasons for individuals to have their genomes sequenced now, as opposed to waiting for improved technology.

Genetics and How We Think About Ourselves

Four of the most important *facts* presented in Part I are repeated here:

- The information stored within the DNA molecules that comprise our genomes greatly affects our appearance, health, and abilities.
- There is no such thing as a genetically perfect person.
- People are not equally fortunate in the genetic lottery of life.
- No one has any appreciable control over their own genome.

The heritabilities of health problems and other human traits vary over a wide range (Table 4.1), but are nearly always substantial. All people are mutants, and all people carry many DNA variants that increase our risk for disease and disability. Although there are no genetically perfect people, this does not mean that our genomes are equal. The reality is that some people have genomes that allow them to thrive and excel, while others have genomes that make them sick or disabled. Finally, no one has any control over their own genome. Just like a person doesn't get to choose their relatives, so no one has any choice about their genome.

I ran for exercise and pleasure nearly all my adult life. In the many road races I entered, I invariably finished somewhere in the middle of the pack—not the worst, but certainly not among the best. When I was a young man, I occasionally thought that if I trained really hard, I could become an elite runner. But I now understand that was just youthful fancy. Yes, if I had trained harder, I could have run faster, but regardless of how much I trained, I never could have become an elite runner. I simply did not have the build, the fluidity of motion, and endurance of the best runners.

Just as my running ability was limited by my genome, so every person's abilities are limited by theirs. Nonheritable factors such as nutrition, injuries, and infections can add or detract from our talents, but ultimately, we are constrained by our genomes. Not everyone is capable of becoming a professional athlete, or a surgeon, or a top business executive. That does not mean, however, that our fates are fixed by our genomes. Within our genetic limits, there is considerable variation in what people actually achieve. We all have known people who are talented but, for various reasons, achieve relatively little in their lives. And we have also all known people who are not particularly talented but nevertheless achieve a great deal. Determination, courage, industry, and other such virtues are very real and greatly affect what we accomplish in life.

Precision Medicine

A great deal of interest and excitement about precision medicine has arisen in recent years (1–3). One of my favorite quotes on this topic comes from Matt Hancock, a former senior minister in the United Kingdom (UK) government who was responsible for the National Health Service (NHS):

> My ambition is that eventually every child will be able to receive whole genome sequencing. . . . We will give every child the best possible start in life by ensuring they get the best possible medical care as soon as they enter the world. Predictive, preventative, personalized healthcare—that is

the future of the NHS—and whole genome sequencing and genomics is going to play a huge part in that. (2019)

Although progress has certainly been slower than I would like, many nations are now advancing toward Mr. Hancock's ambition. Tens of millions of people worldwide have now had at least significant portions of their genomes sequenced (4–6). Research projects involving the genome sequencing of hundreds of thousands of newborns are underway (7–8). In the UK, already 500,000 volunteers (about 1% of the population) have had their genomes sequenced in the UK Biobank project (9), and the NHS recently announced plans to begin sequencing the genomes of all newborns by 2035 (10). Singapore plans to complete genome sequencing on 10% of the nation's population by 2031 (11–12). Researchers in Finland have already generated substantial genome data on 10% of the population and combined it with clinical data (13–14).

My Vision

In Chapter 6, I described how germline DNA testing, and genome sequencing in particular, will substantially improve disease diagnosis and management, drug prescription, disease prevention, and reproductive planning, among other applications. This improvement has, I think, been one of the greatest achievements to date of the science of human genetics. It is a great victory over disease and disability—one that I think we should all celebrate.

I'll conclude Part I with my vision for the future of clinical genetics and my recommendations for people regarding their genomes.

A baby born in the year 2040 is shown in Figure 8.1, upper left. At birth, after the baby's umbilical cord is cut, a small volume of blood is collected from the cord. Since the cord has already been cut, no possible harm or pain can come to the baby from the blood draw. The cord blood cells contain DNA from the baby, not the mother. Most of the DNA extracted from the cells is banked for future testing, but a portion is used for genome sequencing. If the baby has a health problem at birth or is likely to develop one during the early years of life, the sequence information is used to assist with diagnosis and disease management.

Figure 8.1 My vision for the future of clinical genetics

The baby at seven years of age is shown in Figure 8.1, upper right. Her DNA sequence has been stored in her electronic health record. Because geneticists' ability to interpret genome sequences is steadily improving, her sequence is reinterpreted periodically throughout her life. DNA sequence information from her parents and other relatives is used to assist in the interpretation. All the new variants present in the girl that arose through mutation are evaluated.

In Figure 8.1, lower left, the girl is now a young woman. She is thinking about getting married and starting a family. The sequence information generated at birth helps her with reproductive planning and hopefully in choosing a spouse (Chapter 11). Also, throughout her life, the DNA information is used to inform drug prescription, making sure that, as needed, she receives the safest, most effective pharmaceutical at the correct initial dose.

As the years go by (Figure 8.1, lower right), the woman's healthcare providers use the DNA sequence information plus nonheritable factors to predict her risk for cancer and many other later-onset disorders. Enhanced screening and other preventive measures are implemented for any disorders for which she is at especially high risk.

- - -

During my tenure as the head of a germline clinical DNA testing lab, I oversaw the testing of about 500,000 patients. I observed firsthand, over and over, the value to the patients of DNA testing. Based on this experience, I'm absolutely certain that DNA testing will substantially improve people's health and lives.

Here, then, are my recommendations regarding genetics and healthcare:

- Bank your DNA.
- Have your genome sequenced as soon as practical.
- Treat your genome sequence as potent, private information, but do not be afraid of your genome. Danger comes from genetic ignorance, not genetic knowledge.
- Demand that your providers use your genome sequence to improve your healthcare.
- Make sure your genome sequence is retained indefinitely via secure data storage.
- Be willing to share your genetic information anonymously to benefit family members and all others.

No one can hide from genetics. Our genomes affect our lives profoundly whether we like it or not. The real choice is whether to remain ignorant about our genomes or to learn about them and use this information to improve our lives.

PART II:

THE GENE POOL

Chapter 9

INTRODUCTION TO PART II

In 1958, the distinguished medical geneticist James Neel wrote (1):

> We are all familiar with the complex apparatus of modern medicine, which briskly and efficiently swings into action when a patient with congenital heart disease enters hospital for surgery, or a patient with acute renal shutdown comes in for dialysis. But, in our concern for the individual, have we forgotten to set up the team which has as its concern the species as a whole?

Part II of this book is about the genetics of the human species as a whole.

I define the gene pool as the sum of all the genomes of all people alive at any particular time. National and local gene pools can certainly be considered, but in this book, I deal mostly with the gene pool of all living people.

Like Part I of this book, Part II shows how human genetics can be used to improve people's lives. But Part II also covers some topics, such as reproductive planning and the gene pool problem, that are more sensitive and difficult than those covered in Part I. I wrote about these matters not to provoke, but rather because I believe the status of the gene pool is one of the most important issues facing humanity today (2).

Also like Part I, Part II is about the *science* of human genetics. As I explained in the first chapter, scientists cannot control how nature works, but rather try to learn about nature and then communicate this knowledge to others. My *scientific* analysis of the human gene pool has led me to conclude that there is a problem, and that it is essential that we devise solutions.

People sometimes wonder whether humans are still evolving. The answer is absolutely, positively yes. The gene pool is constantly changing through the processes of mutation, random drift, and natural selection. DNA sequence variants are constantly arising, changing in frequency, and vanishing from the gene pool.

I begin Part II with some basic biology about human reproduction and a fairly detailed discussion of reproductive planning. I then proceed to cover the genetics of intelligence—not because intelligence is the only important trait, but rather because it is the one trait that sets humans apart from all other species. I then tackle the difficult and speculative subject of the evolution of human intelligence. Over about the last seven million years, humans evolved from relatively dull, ape-like creatures to a species that has built computers and spacecraft. No one, certainly not myself, knows exactly how this occurred, but I think it's useful to at least try to identify some of the major forces that were involved. Next, I make my best effort to describe how these evolutionary forces have changed in relatively recent times. Through this analysis, I identify the gene pool problem. Finally, I recount some events in the history of gene pool management, and offer some possible solutions to the gene pool problem—solutions that do not resort to racism or elitism.

Chapter 10

PRENATAL DEVELOPMENT AND PRENATAL DNA TESTING

This brief chapter provides some basic information about prenatal development and prenatal DNA testing, which serves as an introduction to the next chapter on reproductive planning.

Overview of Human Prenatal Development

Prenatal development begins with the fusion of an egg cell from the mother's ovaries with a sperm cell from the father's testes. Biologists call this union fertilization (also sometimes called conception). One or two egg cells are usually released from the ovaries during each menstrual cycle. The egg cells are captured into the fallopian tubes, and travel over a period of days through the tubes into the uterus (also called womb) (Figure 10.1). Sperm cells travel through the cervix and uterus into the fallopian tubes where fertilization usually takes place. The fertilized egg cell begins dividing immediately; the dividing cells are now called an embryo.

By day 5 post fertilization, the embryo consists of a ball of over 100 cells (1). At this stage, the embryo has already split into two groups of cells. One group develops into the baby, while the other group develops into the extraembryonic tissues (see below). At days 6–12 post fertilization, the embryo reaches the uterus and implants into the uterine wall. Formation of the nascent structures of the body then proceeds rapidly. At about day 16 post fertilization, the cells that will develop into the child's egg or sperm cells split off from the other cells of the embryo (2–4). Mutations that occur in these precursors to the egg and sperm cells may become germline mutations in the child's children. By about day 22 post fertilization, the rudimentary heart begins beating. By day 56, most of the tissues and organs of the body can be recognized. At about this point, the developing baby is no longer called an embryo, but begins to be called a fetus.

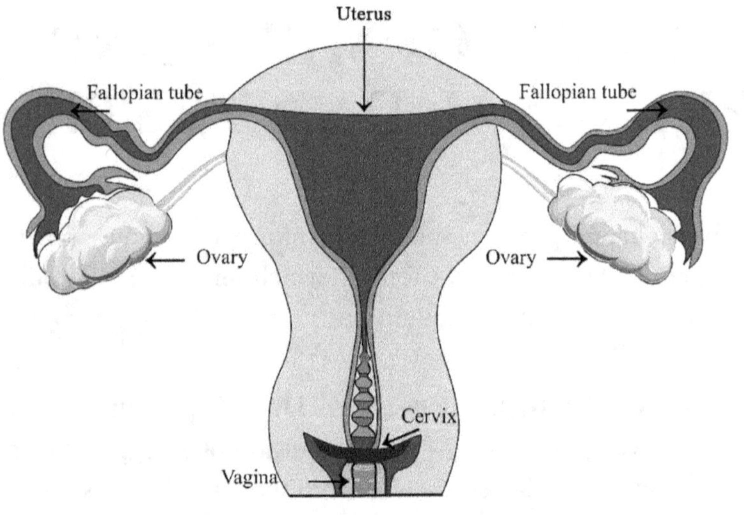

Figure 10.1 Female reproductive system

Pregnancy timing is calculated from the first day of the mother's last menstrual period, about two weeks before fertilization. A pregnancy normally lasts about 40 weeks (9 months) and is typically split into three trimesters: the first trimester (weeks 1–12), the second trimester (weeks 13–27), and the third trimester (weeks 28–40). Note that weeks post fertilization are about two weeks less than weeks of pregnancy.

As mentioned above, a portion of the cells from the early embryo develop not into the baby, but rather into the extraembryonic tissues: the amnion, chorion, placenta, and umbilical cord (Figure 10.2). The amnion forms an inner membrane that encloses the fetus in a protective fluid. The chorion forms an outer membrane that also encloses the fetus and that extends into the uterine wall (chorionic villi). The placenta, which develops from the chorion, serves as an intermediate organ between the mother and the baby. Oxygen, nutrients, and other molecules pass from the blood vessels in the mother's uterus to vessels in the placenta and then on to the fetus though the umbilical cord.

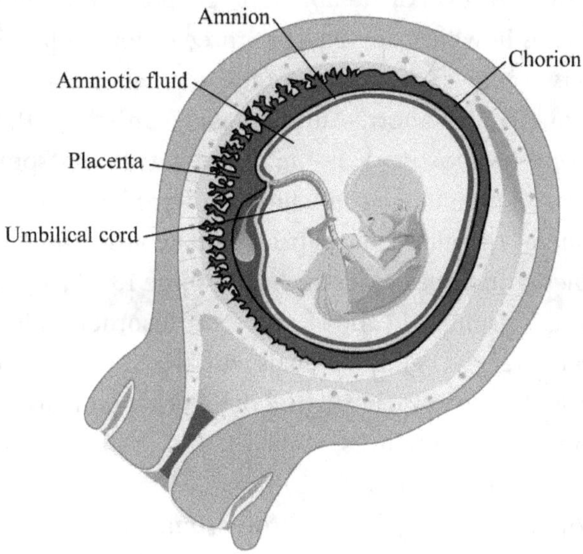

Figure 10.2 Human fetus at about 15 weeks post fertilization

During normal birth, the membranes rupture, releasing fluid, and the baby is delivered first, followed by the placenta. After the umbilical cord is cut, blood can be easily drawn from the cord without any possible harm to the baby. Importantly, because cells in the placenta, chorion, amnion, cord, and cord blood developed from the early embryo, they have the DNA sequences of the baby, not the mother.

Many things can and often do go wrong during prenatal development. Many pregnancies naturally terminate in what is called a spontaneous abortion (also called a miscarriage if it occurs before about 20 weeks, or a stillbirth after about 20 weeks). Spontaneous abortions are estimated to occur in 40%–75% of all pregnancies and about 15% of recognized pregnancies (5–8). Often the spontaneous abortion occurs so early in development that the woman does not realize that she is pregnant. The frequency of spontaneous abortion drops as pregnancies proceed. There are many causes of spontaneous abortion, but certainly a frequent cause—and probably the most frequent—is aneuploidy, or an abnormal number of chromosomes in the developing baby (Chapter 3) (6, 9–11).

Only a small fraction of aneuploid pregnancies result in live births. All pregnancies in which the fetus has three or four copies (instead of the normal two copies) of *every* chromosome, called triploidy or tetraploidy, spontaneously abort. All pregnancies in which the fetus has only one copy (monosomy) of *one* of the autosomes spontaneously abort. Nearly all pregnancies in which the fetus has three copies (trisomy) of *one* of the autosomes also spontaneously abort (6, 12–14). Rarely, babies with three copies of chromosome 13, 18, or especially 21 are born alive. Trisomies 13 and 18 are severe disorders; affected babies usually die in infancy (14). Individuals with trisomy 21, however, often survive these days into their 50s and 60s. Although many individuals with trisomy 21 live long lives, about 80% of trisomy 21 pregnancies end in spontaneous abortion (6, 12).

Aneuploidies involving the sex chromosomes have less severe health effects than those involving the autosomes. For example, although they do have health problems, females with only one X chromosome (Turner syndrome) or with three X chromosomes are relatively healthy. Similarly, males with one Y and two X chromosomes (Klinefelter syndrome), or one X and two Y chromosomes are also relatively healthy. About 99% of female pregnancies with one X chromosome spontaneously abort (6, 12). Why only a tiny fraction of monosomy X pregnancies progress to live birth is not completely understood. However, one reason appears to be that some living monosomy X individuals are mosaics—that is, they have one copy of X in only a portion of their cells, with the normal two copies in the remainder of their cells (15).

Most germline mutations occur in fathers (Chapter 3). Aneuploidies, however, are an exception; most occur in mothers. The frequency of both aneuploid pregnancies and aneuploid births increases sharply with advancing maternal age (8–10, 16) (Figure 10.3).

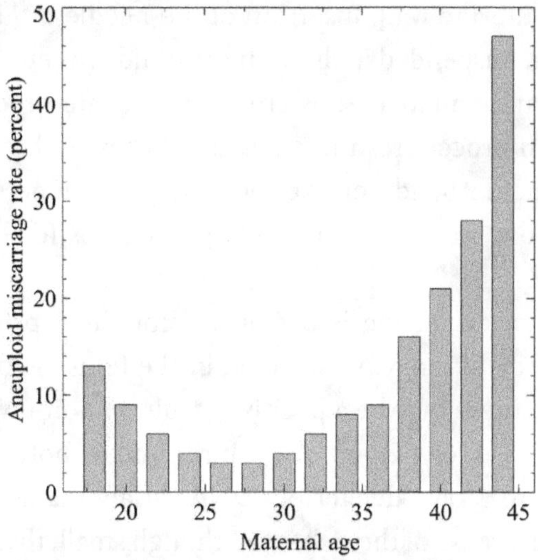

Figure 10.3 Rates of miscarriage due to aneuploidy versus maternal age
(adapted from reference 8, Figure 1)

Invasive Prenatal Testing

Prenatal testing means testing of DNA, or other biomolecules, from a fetus before birth. Invasive testing means that a small tissue specimen needs to be collected from the fetus. For most invasive prenatal testing, one of two types of tissue are collected: either chorionic villi or amniotic fluid (Figure 10.2). Chorionic villus sampling (CVS) is typically performed between 10 and 13 weeks of pregnancy (17) (descriptions of CVS for patients can easily be found by searching the internet). Chorionic villi—projections of the chorion into the uterine wall—are collected either by inserting a thin tube through the vagina and cervix or by inserting a needle through the abdomen. In either approach, ultrasound is used to guide the procedure.

Amniocentesis is usually performed between 15 and 20 weeks of pregnancy (18) (descriptions of amniocentesis for patients can easily be found by searching the internet). In amniocentesis, a small volume of the amniotic fluid that surrounds the developing fetus within the amniotic sac is collected by inserting a long needle through the mother's abdomen and uterine wall. Insertion of the needle is guided by ultrasound imaging. Suspended in the amniotic fluid are cells from the fetus.

Both CVS and amniocentesis are generally safe procedures. However, with both procedures there is a small chance that the fetus may be harmed. In the hands of experienced practitioners, the current risks of fetal loss are about 1 per 500 procedures for both CVS and amniocentesis.

Invasive prenatal testing is performed for three primary reasons. First, an abnormality may be detected in the fetus via ultrasound imaging. Second, there may be a positive result on non-invasive prenatal screening (NIPS) (see below). And third, one or both of the parents may be known to carry a deleterious DNA variant that is relevant to the baby's health. Because of the risk, even though small, that the fetus may be harmed, invasive testing is rarely performed today on apparently healthy fetuses, although this might change in future (19).

Fetal DNA from an invasive procedure can be tested using the full range of available DNA laboratory tests. Sometimes just one nucleotide at a single chromosomal position is tested. Sometimes a panel of genes is sequenced, or even the entire fetal genome may be sequenced. When prenatal testing is indicated by an abnormal ultrasound, it's common for the DNA of both parents and the fetus (a trio) to be tested concurrently. This is done so that new deleterious variants in the fetus that arise through mutation can be identified (see Case 2 in Chapter 3).

Depending upon the type of prenatal test performed and the amount of fetal tissue collected, test results are available anywhere from a few days up to about three weeks after the procedure. Parents will therefore receive the results from invasive testing anywhere from about 12 to 23 weeks of pregnancy.

Many parents elect abortion if the DNA results indicate that the baby will have a significant health problem, but many others do not. It's not unusual for parents to opt for prenatal testing to prepare for the birth and care of an affected child. At least where abortion is legal, the choices are always made by the parents. Genetic counselors and other healthcare providers are trained to be, and I think nearly always are, *nondirective* in their conversations with the parents. They provide the parents with as much relevant information as possible, but then leave decisions solely to the parents. I don't know of any laws or regulations in any country that require an affected fetus to be aborted.

Elective abortion is clearly a sensitive and controversial subject. National and regional statutes regarding elective abortion vary widely (20–21). Some nations and regions ban abortion in all or almost all circumstances. Many others permit abortions, but only up to a certain point in the pregnancy. The situation is in flux, but as of the time of this writing there are states in the US like Texas and North Dakota that ban abortion entirely, states like Georgia and Florida that ban abortion after 6 weeks of pregnancy, states like North Carolina and Nebraska that ban abortion after 12 weeks, and states like Ohio and Massachusetts that permit abortions up to 22–24 weeks. There is also variation in whether the abortion limits apply to weeks of pregnancy or weeks post fertilization. Clearly, in nations or regions that ban abortions after 12 weeks of pregnancy or less, invasive prenatal testing cannot be performed early enough for elective abortion. Also, only a small fraction (I *estimate* this fraction in the US to be 2%) of total elective abortions are performed because of a health problem in the fetus. Nearly all are performed because the pregnancy was unintended.

Non-Invasive Prenatal Screening (NIPS)

In 1997, Dennis Lo and colleagues reported that the blood of a pregnant woman contains extracellular ("cell-free") DNA from the fetus (22–23). This discovery led to the rapid development of maternal blood tests to screen for DNA abnormalities in the fetus (24–25). These tests are called non-invasive prenatal screening (NIPS) (also sometimes called non-invasive prenatal testing, or NIPT). The primary advantage of NIPS over invasive tests is that there is no risk to the fetus. NIPS is also less expensive than invasive DNA testing.

NIPS is methodologically complicated by the fact that the cell-free DNA within a pregnant woman's blood contains more DNA from the mother than from the fetus. The fraction of cell-free DNA from the fetus ranges from 0% at the start of pregnancy up to about 30% in the third trimester (25–26). Despite this obstacle, careful analysis of the test results usually allows detection of abnormalities in the fetal DNA. Today, the primary application of NIPS is to detect aneuploidies.

NIPS for detection of aneuploidies is highly accurate (27–28). False results do rarely occur, however, particularly with twin pregnancies and when the mother is obese. In addition, because the "fetal" cell-free DNA in the mother's blood comes from the placenta and not from the fetus proper (29), there is a chance that the placenta is aneuploid, but the fetus has normal chromosome numbers. A positive NIPS result should therefore always be confirmed by an invasive DNA test, usually amniocentesis.

NIPS is typically performed at 11–12 weeks of pregnancy (24–25, 30). This allows time for the confirmatory amniocentesis to be completed before the end of the second trimester. Although, as described above, aneuploid pregnancies are common, the great majority of aneuploid pregnancies spontaneous abort earlier than 11 weeks. Therefore, only a small fraction (roughly 1%) of NIPS tests for aneuploidy are positive (27).

Although the primary purpose of NIPS today is to detect aneuploidy, researchers have demonstrated that cell-free DNA from the mother's blood can be used to test for single-gene disorders and may even be used to perform full fetal exome or genome sequencing (30–36). These other applications of NIPS have not yet been widely adopted, but could be in future.

NIPS is now widespread throughout the developed world (24, 37). It is generally offered to all pregnant women, although it is particularly recommended to older mothers. NIPS has resulted in a substantial decrease in the number of invasive prenatal tests (25). It has also resulted in a substantial decrease in the number of aneuploid babies born, although the rate of elective abortion of aneuploid fetuses varies greatly among nations and regions (24, 38–39).

Summary

- Prenatal development is a complex 38-week process that begins with a fertilized egg cell and ends with the birth of an infant.

- Some of the cells from division of the fertilized egg cell develop into the baby and others develop into the extraembryonic tissues: the amnion, chorion, placenta, and umbilical cord.
- Many things can and often do go wrong during prenatal development. Roughly 60% of all pregnancies and 15% of recognized pregnancies spontaneously abort.
- The most frequent known cause of spontaneous abortion is aneuploidy, or an abnormal number of chromosomes. Nearly all aneuploid embryos and fetuses spontaneously abort.
- DNA from the developing fetus may be tested through both invasive and non-invasive procedures.
- Invasive testing is practiced on only a small fraction of pregnancies and nearly always involves either chorionic villus sampling (typically performed at 10–13 weeks of pregnancy) or amniotic fluid collection (typically performed at 15–20 weeks).
- All types of clinical DNA tests may be performed on fetal DNA obtained from the invasive procedures.
- Non-invasive prenatal screening (NIPS) is a maternal blood test used to assay the fetal DNA. Today, NIPS is used primarily to test for aneuploidies, but in the future may be used to test for other genetic abnormalities.
- In developed countries, NIPS for aneuploidy detection is currently performed on a substantial fraction of all pregnancies.

Chapter 11

GENETIC REPRODUCTIVE PLANNING

Impact of Single-Gene Disorders

As presented in Chapter 4, each single-gene disorder is caused by one or two highly penetrant, deleterious sequence variants. Deleterious variants in about 5,000 different genes are known today to cause single-gene disorders (1). Although individually each of these disorders is rare (Table 4.3), because there are so many of them, they collectively make a substantial impact on human health.

It is estimated that 2%–6% of people, or approximately 350 million people worldwide, are affected with single-gene disorders (2–4). Single-gene disorders comprise at least half of what are known as rare diseases. The definition of rare disease varies from country to country (2, 5), but the term often refers to a disease that affects less than 1 in 2,000 people.

Average healthcare costs are much higher for those with single-gene disorders or other rare disease compared to those who are unaffected (2–4, 6–12). A study in Western Australia concluded that individuals with a rare disease comprised 2% of the population yet consumed 5%–10% of the inpatient hospital costs (9). A group of Canadian investigators reported that patients with genetic disorders incurred healthcare costs per patient that were 5–20 times those for the general population (10). In the US, children with genetic disease were reported to account for between 11% and 46% of total pediatric healthcare costs in 2012 (11). And also in the US, a 2016 analysis of healthcare expenses found that children and adults with rare diseases accounted for nearly half the national total (12).

In addition to the direct healthcare costs of single-gene disorders, there are also indirect costs to the caregivers, often parents. Caregivers frequently see negative impacts on their employment, finances, health, and relationships (13–16). Economists who have tried to quantify these indirect costs have reported that they amount to 14% (6) and 21% (17) of the total cost of the rare disease.

Although treatments for single-gene disorders are steadily improving, many of these treatments are exceptionally expensive. For instance, the newer drugs mentioned in Chapter 6 for treating cystic fibrosis typically cost over $250,000 per year (18–20). Gene therapy is another promising avenue for treating patients with single-gene disorders (21–22). Gene therapies, however, are also extraordinarily expensive, typically costing $1 million or more per treatment (23–24). Although the new treatments for single-gene disorders are wonderful advances for patients, there is no denying their enormous cost. And in most cases, the new treatments are not completely effective; they improve the patient's condition, but don't entirely cure the disease.

Nearly all parents would like to avoid having a child with a significant health problem or disability. Fortunately, current genetic knowledge and technology allow this to be achieved in many cases through a process called reproductive planning. Medical geneticist Aubrey Milunsky succinctly described reproductive planning as "choices not chances" (25).

Reproductive planning is expanding steadily throughout the world. Currently, it mainly covers aneuploidies and single-gene disorders, but genetically complex disorders may be included in the future. Aneuploidies are mostly avoided through NIPS (previous chapter). Reproductive prevention of single-gene disorders is the focus of this chapter.

Assisted Reproductive Technologies (ARTs)

Infertility is common. Approximately 15% of couples fail to achieve pregnancy after 12 months of regular unprotected intercourse. The cause of infertility is about equally likely in the man or woman. Primarily in an effort to help infertile couples, a number of new reproductive technologies have been developed, most over the last few decades. Collectively, these are termed assisted reproductive technologies (ARTs). Rather than trying to describe all procedures and practices that have been developed, I cover only the primary ARTs. Note also that since ARTs are changing rapidly, this section will probably become outdated more rapidly than other portions of the book.

Although most do not, I include artificial insemination (AI) (also called intrauterine insemination) as one of the primary ARTs. AI is the earliest and simplest type of ART (26–27). It has been known for at least 130 years that donated sperm can be used to impregnate women. Sperm from a donor is simply injected, at the appropriate point in her menstrual cycle, into the uterus of a woman who desires to become pregnant. Sperm from donors is often frozen within large commercial sperm banks. Women pore through catalogues to choose their donor based on characteristics such as appearance, education, health, and occupation. Tall, handsome, and highly educated men are favored. Sperm donors are paid, and these days, donors are usually tested for at least some genetic abnormalities. Although widely available, AI is still responsible for only a small fraction of live births. It was reported in one study that 0.7% (1 in 140) of American women of reproductive age had used donated sperm in the period 2015–2017 (28).

AI was previously used primarily by couples in which the woman was fertile, but her husband was not. Today in the US, AI is used mainly by lesbian couples and single women. Rarely, AI is used by heterosexual couples in which the man carries a dominant deleterious variant for a single-gene disorder; in these cases, sperm from another man who does not have such a variant will then be used to impregnate the woman. Also rarely, when the man is fertile but the woman is not, or when the woman carries a dominant deleterious variant, a surrogate mother (also called a gestational carrier) will receive AI using the man's sperm. For a fee, the surrogate will carry the baby to term and deliver. By legally binding contract, the baby is then turned over to the man and his partner. The current going rate for surrogate mothers in the US is about $50,000 per pregnancy.

Fertile women produce about 400 mature egg cells during their lifetimes. Fertile men produce roughly 200 million sperm per ejaculate. Therefore, women can have only a limited number of offspring during their lifetimes, while men can have an essentially unlimited number of children. In principle, through ARTs, one man could easily biologically father all 3.6 million babies born each year in the US.

- - -

Another primary ART is in vitro fertilization (IVF) (29). IVF is defined as fertilization of egg cells outside of the body. Mature egg cells are collected from a female donor. Under normal conditions, only one or two egg cells are released from the ovaries per menstrual cycle. To stimulate the maturation of larger numbers of egg cells, the woman receives multiple hormone injections. At the appropriate time, egg cells are harvested from the ovaries by a minor surgical procedure. In most cases, 10–20 egg cells are harvested per cycle. These egg cells may be immediately fertilized or frozen for future use (30).

Originally, the harvested egg cells were fertilized by simply mixing an egg cell with sperm cells in a small plastic container, but more recently, intracytoplasmic sperm injection (ICSI) has been employed. In ICSI, a single sperm cell is injected directly into an egg cell. ICSI was originally utilized in cases where the man was subfertile, with low sperm count or abnormal sperm. Currently, however, in many IVF labs, ICSI is the preferred method of fertilization (31–32).

After fertilization, the resulting embryos are cultured for three to six days in incubators before being transferred into the mother's uterus. During this culturing, cell division occurs; the transferred embryos often contain over 100 cells. Unused embryos are usually frozen so that they are available for future attempts at pregnancy. Donated egg cells, sperm cells, and/or embryos, as well as surrogate mothers (33), may be used in IVF.

Sometimes, DNA from the embryos is tested prior to transfer into the uterus. This process is called preimplantation genetic testing (PGT) (34). In PGT, one or more cells are removed from the embryo, and DNA is extracted from the cells and tested. These days, the cells removed from the embryo are usually those that give rise not to the fetus, but rather to the extraembryonic tissues (Chapter 10). Usually, removal of the cells does not appear to harm the embryo. Although PGT is technically demanding because of the small amounts of DNA obtained, virtually all types of clinical DNA tests can be performed on the material. PGT is often performed when the mother is older (because of the increased risk of aneuploidy with maternal age), or when the mother and/or father carry variants that cause single-gene disorders. Only embryos that are expected to lead to children without genetic disorders are then transferred.

The process of IVF is rough on the woman and is quite expensive. The woman receives a number of painful hormone injections and must endure unpleasant procedures. And since pregnancy is often not achieved on the first or second try, she may have to undergo the process multiple times. The overall rate of successful pregnancy through IVF in the US in 2021 was approximately 30% per cycle, although this rate strongly depends upon a number of factors, especially the age of the mother (35–36). At the time of writing, the cost in the US of a single IVF cycle is on the order of $25,000 (this price does not include costs of purchasing donated eggs, sperm, or embryos, freezing of eggs or embryos, or PGT). Despite the difficulties and the high cost, the desire of many people to have children is so strong that IVF is extensively utilized. About 2.5% of newborns in the US today result from pregnancies achieved using IVF (35). Worldwide, about 10 million people (roughly 1 out of every 1,000) were conceived using IVF (34, 37).

Are ARTs safe? The short answer is yes. Millions of healthy babies have been born using ARTs. However, there are reports of slightly increased rates of health problems in children born using ARTs compared to children who were naturally conceived (37–39). And since ARTs are relatively new, it is not yet known whether they will have long-term effects on the health of the offspring. At the time of writing, the oldest individuals born via IVF are only about 45.

Options for Couples at Risk of Having a Child with a Recessive Single-Gene Disorder

Case 8: Reproductive Planning for Recessive Disease
Jolanda and Eric were a young married couple who very much wanted children. Because they were both of African ancestry, they knew they had a much higher risk than most couples of having a child with sickle cell anemia. This they wanted to avoid. What could they do?

To start, Jolanda and Eric could have their own DNA tested. Sickle cell anemia is a recessive disease—both Jolanda and Eric needed to be carriers for them to have an affected child (Figure 4.4C). They each had a roughly 8% chance of being a carrier for the sickle cell variant, and there was a 0.6% chance (about 1 in 150 African American couples) that they would both be carriers. Despite this small chance, they decided that the consequences of having an affected child outweighed the trouble and cost of DNA testing.

Jolanda and Eric made an appointment with a genetic counselor. The counselor told them that they had a few options for testing. They could be tested for only the sickle cell variant; they could have what is called a carrier test, in which hundreds of different genes involved in recessive disorders, including the sickle cell gene (*HBB*), are sequenced; or they could have whole genome sequencing, which would provide them with the maximum amount of genetic information about reproduction, as well as much other information that could be used to improve their own healthcare. After carefully considering the matter, they decided to go for the works: genome sequencing. A small

> vial of blood was drawn from each and sent to a lab for the testing.
>
> In three weeks, they returned to the counselor to receive the test results. They were relieved to learn that neither carried the sickle cell variant. However, they also learned that Jolanda carried recessive, deleterious variants in five different genes, and Eric had recessive, deleterious variants in three genes. But since the couple did not have deleterious variants in the *same* gene, they were at low risk for having a child with a recessive disease. This was great news. They were also happy that their DNA sequences would now be available to help with their own drug prescription and disease prevention.

Fortunately for Jolanda and Eric, they were not at high risk for having a child with a recessive disease. However, for couples in which both the man and woman are heterozygous for deleterious variants in the *same* gene, there are three primary options for avoiding the birth of an affected child. These options are listed in Table 11.1 and described below.

Table 11.1
Options for couples who are heterozygous for deleterious variants in the same recessive gene

- Do not reproduce together
- Conceive naturally, followed by invasive prenatal testing and elective abortion in the case of an affected fetus
- Conceive using in vitro fertilization (IVF) coupled with preimplantation genetic testing (PGT)

First, the couple could decide not to have children together. They could go childless, adopt, or split up and find different reproductive partners. They could also choose to use a sperm, egg, or embryo donor. Second, the couple could become pregnant, and then opt for invasive prenatal testing. If the fetus is affected, it could be aborted. The chance of an affected child is 25% with each pregnancy (Figure 4.4C). Third, the couple could opt for IVF coupled with PGT. Only embryos that do not have both of the deleterious variants would be transferred into the mother. Note that of these three options, only the second involves elective abortion.

It's important to understand that reproductive planning is not just for healthy individuals. In many cases people affected with genetic disorders can, with care, have healthy children. For example, if a person with a single-gene recessive disorder reproduces together with a person who does not carry a deleterious variant in the same gene, then all the offspring of that couple will be unaffected carriers.

Options for Couples at Risk of Having a Child with an Inherited Dominant Disorder

Case 9: Reproductive Planning for Dominant Disease
John was a 24-year-old man with a family history of Huntington's disease, a dominant, deadly, late-onset neurological disorder (Chapter 3). John's maternal grandmother and uncle died of the disease, and his mother was beginning to show symptoms. John was therefore likely at 50% risk for developing Huntington's. John and his wife, Sandy, wanted to have children, but they didn't want to have a child who would develop Huntington's. What could they do?

Fortunately, John was open with Sandy about his family history even before they were married. If he had not been open, the news would have been a shock to her. Even worse, she could have learned of John's family history after they had children.

The first reproductive planning decision for John and Sandy was whether John should receive DNA testing to determine whether he would develop Huntington's. This was obviously a very difficult decision. Some in John's position decide to be tested; others do not. After careful consideration, including genetic counseling, John decided to proceed with the testing. Unfortunately, John was found to be

> heterozygous for an expanded run of repeats in the *HTT* gene and therefore would develop Huntington's disease later in life (Table 3.2).
>
> After digesting this information, John and Sandy decided that they still wanted to have a child. They opted for IVF coupled with PGT. The first cycle was unsuccessful, but Sandy became pregnant in the second cycle. About eight months later, she delivered a healthy baby boy who would not develop Huntington's. They named him John.

The options for couples in which either the man or woman carries a deleterious variant in a dominant gene are much the same as for recessive disease (Table 11.1). The couple could decide not to reproduce together; they could become pregnant, test the fetus, and elect abortion if it is affected; or they could choose IVF with PGT.

Preconception Reproductive Planning

Although most reproductive planning today occurs *after* the woman is pregnant, I think a much better approach is to begin planning *before* the woman becomes pregnant. One option for preconception reproductive planning is to choose a reproductive partner based in part on the partner's genome. This approach may sound strange or unrealistic, but as you'll see, is already being effectively implemented.

Rabbi Yosef Ekstein and his wife lost four children to Tay-Sachs disease. This recessive disease is caused by defects in the *HEXA* gene, which codes for an enzyme that catalyzes the breakdown of a specific type of lipid (40). In the absence of this enzyme, levels of this lipid build up to toxic levels in the body. Tay-Sachs is a severe disease. There is currently no cure, and affected children usually don't live more than five years. In the past, Tay-Sachs was relatively common in the Ashkenazi Jewish population due to what geneticists call founder variants (Chapter 13).

Rabbi Ekstein didn't want other couples to suffer like he and his wife did, so he started a program called Dor Yeshorim, a Hebrew phrase that roughly translates as "righteous generations" (41–42). Here's how Dor

Yeshorim works. Young people who want to someday get married have their DNA tested for the Tay-Sachs founder variants and a number of other Jewish deleterious founder variants. The current standard testing panel covers more than 50 single-gene disorders. When a couple is thinking about possible marriage, they query the Dor Yeshorim database to learn if they are carriers for deleterious variants in the same gene. Nearly all of the small fraction of couples who are genetically incompatible in this way do not get married, but rather find different partners.

Dor Yeshorim has been a resounding success. Without resorting to abortion, the birth rate of Tay-Sachs disease in many Jewish communities has dropped to near zero (40). The following is an excerpt from a letter written in 2008 by the director of pediatrics at Kingsbrook Jewish Medical Center to Rabbi Ekstein:

> We met many years ago and I just wanted to write to you a final chapter regarding Tay-Sachs disease. As you know Kingsbrook Jewish Medical Center was the largest Tay-Sachs Center for the United States. Each Tay-Sachs patient who was admitted to the hospital for Long Term Care would cost Medicaid at least $130,000 per year. Through your aggressive genetic testing program to prevent this serious medical problem we have eliminated this disease. Our last Tay-Sachs patient passed away five years ago and there have been no admissions since.

Geneticists have now accumulated sufficient knowledge and technology to greatly reduce the incidence of thousands of recessive diseases. Preconception reproductive planning would rule out only a small number of potential reproductive partners. Nearly all people carry a few highly penetrant deleterious variants, mostly for recessive disease (see Chapter 5 and Case 8 above), but the probability that two individuals picked at random from the human population have deleterious variants in the same gene is low—on the order of 1% (43–45). Therefore, 99 out of 100 individuals would be genetically acceptable as a reproductive partner for someone who wants to avoid a child with recessive disease.

However, if the couple are both from the same reproductively isolated group, such as a small island population or a religiously isolated group like Ashkenazi Jews, or if the couple are from a society where marriages between cousins are common, then the probability of an incompatible partner for recessive disease increases substantially (44–45). For first cousins, it is about 15% (1 in 7 couples) (Figure 15.1) (43).

But what about preconception planning for dominant disorders? With dominant disorders, the situation is more difficult. A person who is heterozygous for an autosomal dominant or X-linked deleterious variant cannot eliminate their risk of having an affected child by choosing a different partner. My strong opinion is that anyone in this situation should discuss the matter openly with their prospective reproductive partner before marriage/conception (Case 9 above). Yes, such a discussion could cause the couple to break up, but I think it is much better to be open at the start than to have the truth come out later. Some readers may conclude that this will make such individuals unmarriageable, but I disagree. None of us is perfect; we all have many flaws. Yet most people still manage to find partners. And as I described above, there are other options to avoid the birth of a child who is affected.

Complex Disorders

To date, little reproductive planning has been devoted to genetically complex disorders, but that may change in the future. As risk indices (Chapters 4 and 6) improve, it will gradually become possible for a prospective reproductive couple to learn their chances of having a child affected with a genetically complex disorder. These risks will only be imprecise estimates, but for some complex disorders, like type 1 diabetes and Alzheimer's disease, the risks calculated in this way may be significant.

Although I don't think consideration of complex disorders would rule out many prospective partners, it would rule out more than when considering only single-gene disorders. Also, for complex disorders, rather than a partner just being acceptable or unacceptable, there would be gradations in genetic compatibility. A prospective reproductive partner might be highly compatible in terms of complex disorders, highly incompatible, or anything in between. I can even envision that someday an online service will, for a fee, take a person's genome sequence, compare it to the genome sequences of many individuals of the opposite sex, and return a list of individuals who are most compatible genetically.

One type of reproductive planning for complex disorders is already available, although it is controversial and rarely used (46–48). When multiple embryos from a couple are generated by IVF and tested via PGT, it is possible to compute risk indices for each embryo and then transfer the embryo with the lowest risk scores. The majority of geneticists today disapprove of this approach (47–48), mainly, I think, because the indices are at such an early stage of development, but the broader public is not so negative (46, 49).

Aneuploidies

Aneuploidy in fetuses can now be detected relatively early in pregnancy by NIPS (Chapter 10). NIPS involves no risk to the fetus and is relatively inexpensive. When parents receive a positive NIPS result for aneuploidy, and this result is confirmed by invasive testing, their only choice to avoid the birth of an affected child is elective abortion.

Elective Abortion

Reproductive planning involves many difficult ethical and moral issues. I won't delve deeply into this topic, but I will include a little information here about elective abortion. As mentioned in the previous chapter, only a small fraction of elective abortions (perhaps 2% in the US) are performed because the fetus has a genetic abnormality. Nearly all are performed because the pregnancy was unintended (50–51).

However, in cases where the fetus is known to carry a genetic abnormality, parents face an obviously difficult and intensely personal decision about elective abortion. I would never presume to try to make such a decision for others. Some of the primary factors couples may take into account when considering a decision about elective abortion are listed in Table 11.2.

Table 11.2
Some factors to consider when faced with a fetus with a genetic abnormality

- Severity of the disorder
- Penetrance of the deleterious variant(s) (many who carry low-penetrance deleterious variants remain unaffected throughout their lives)
- Whether the disorder exhibits variable expressivity (often the severity of a disorder cannot be predicted from DNA sequences alone)
- Financial costs of caring for the affected child
- Time and energy required for caregiving
- Effect on other family members
- Chance that research may lead to effective treatments for the disorder during the lifetime of the child (for instance, a child born today with a causative variant for Huntington's disease might have effective treatment options by the time the child is 40 years old and begins to develop symptoms)

It's unfortunate that such tough decisions fall on the shoulders of young adults. Young adults face many challenges; they are actively establishing relationships, careers, finances, and personal philosophies. It's hard to add the burden of reproductive planning. But we have no choice. Older, more experienced people rarely reproduce.

I'm not a fan of abortion, but I think severe restrictions on elective abortion are unwise. I would never support any requirement that parents abort a fetus with a particular genetic makeup, but I also feel that this option should not be ruled out. I don't in any way blame parents for wanting healthy children. Hopefully, shifting much reproductive planning from postconception to preconception will reduce the demand for *genetic* elective abortion.

Impact of Reproductive Planning

If current trends continue, reproductive planning will steadily expand around the globe. DNA testing for healthcare will catalyze reproductive planning. When virtually everyone has their genome sequenced as a routine part of healthcare, reproductive planning will accelerate.

Because the costs of caring for people with serious genetic disorders are so high, reproductive planning has the potential to substantially reduce healthcare costs (or, perhaps more realistically, allow available healthcare dollars to be redirected to other purposes). Although we are in the very early stages, the incidences of some genetic disorders in some places have already begun to decline through reproductive planning (52–57).

Regarding the gene pool, if reproductive planning becomes widely and consistently applied, it could have an effect on highly penetrant deleterious variants. Population frequencies of deleterious variants for severe dominant disorders like Huntington's disease (see Case 9 above) could drop. Frequencies of variants that cause recessive disease could also decrease, but the rate of decrease would be much slower than for dominant disease (58–59). Frequencies of lower-penetrance deleterious variants, however, are unlikely to be much affected by current forms of reproductive planning.

Finally, it's important to keep in mind that although the incidence of severe disease can be reduced through reproductive planning, no couple can produce a genetically perfect child. Genetically perfect people do not exist.

Summary

- Although single-gene disorders are individually rare, because there are thousands of them, total societal costs of caring for people with single-gene disorders are high. These costs include both direct healthcare costs and indirect costs to the caregivers.
- Assisted reproductive technologies (ARTs) have been developed to help subfertile and infertile couples reproduce and avoid genetic disease. ARTs include artificial insemination (AI), in vitro fertilization (IVF), and preimplantation genetic testing (PGT).

- Through DNA testing, geneticists can now predict a reproductive couple's chance of having a child with a single-gene disorder. This ability may expand in future to include genetically complex disorders.
- Couples can avoid the birth of a child with a genetic disorder via several options, including partner selection, prenatal testing, and ARTs. Abortion can be avoided through partner selection and ARTs.
- Preconception reproductive planning through selecting a genetically compatible partner is an attractive option that will hopefully grow in popularity.
- At least in the near future, genetic reproductive planning will probably have only a limited impact on the gene pool.

Chapter 12

GENETICS OF INTELLIGENCE

Definition and Rationale

For this book, I broadly define intelligence as the ability to solve problems of all types. People who are more intelligent solve problems more rapidly than people who are less intelligent. More intelligent people can also solve difficult problems that less intelligent people cannot solve at all.

Why focus on intelligence? If I had just wanted to describe the genetics of a complex trait, there are plenty to choose from. I chose intelligence because it is the one trait that distinguishes humans from all other species. Humans are the only species that can communicate fluently via speech and written language. Humans are the only species that has developed mathematics. Humans are the only species that has built complex machines. Humans now dominate the globe, not because of our physical attributes, but because of our special level of intelligence.

Few people are upset by discussion of the genetics of, say, height or blood pressure, but intelligence is different. Any discussion of the genetics of intelligence arouses passions and invites criticism and even condemnation. I assume this is because labeling people as less intelligent may result in their suffering even more denigration, and also because it has been claimed that average intelligence differs among human populations. I don't relish delving into such a controversial subject, but I think intelligence is so important to humanity that it needs to be carefully considered. My purpose in writing this chapter was not in any way to demean those with lower intelligence, but rather to accurately and realistically describe what is currently known about the genetics of intelligence.

Variation in Intelligence

There is no perfect way to measure intelligence. We instead need to rely on flawed methods such as intelligence (IQ) tests, school achievement tests, and years of educational attainment. Intelligence tests are culturally biased (1–2). School achievement is greatly influenced by access to quality education. And years of educational attainment depend in part upon financial resources. Nevertheless, when the best measures of intelligence are examined, considerable differences among people are found. Some have IQs of 140; others 70. Some score in the top 10% in their college entrance exams; others in the lowest 10%. Some obtain doctorate degrees, while others do not finish high school.

Imagine a group of 10-year-old schoolchildren. On a few separate days spaced out during the school year, the children are organized into races across the playground to determine who are the fastest runners. Some of the kids will consistently place high in the races. No one is surprised by this. Some people are naturally more athletic than others. Now bring the same group of children into the classroom and give them tests in various subjects over the course of the school year. Again, some will consistently get the highest marks on the tests. This should surprise no one. Some people are naturally more intelligent than others.

As described in the next chapter, our special intelligence evolved during the millions of years since the common human-chimpanzee ancestor. This change could not have occurred unless individuals varied in intelligence. Mutation created the variants that caused higher intelligence in our ancestors, and natural selection and other evolutionary forces led to the increases in population frequencies of these variants.

Importance of Intelligence

Despite the imperfections in the available methods to estimate intelligence, these estimates correlate with many things that we connect to success in life (3–15). Compared to people who score low by these measures, individuals who score high, on average, have higher incomes and largely comprise professions such as medicine, law, and business leadership. They also live longer lives and are more likely to be married and employed. Of course, there are many individuals who are exceptions, but overall, these trends hold firm.

It was only in about the last 6,000 years that human civilizations began to develop. Why didn't our ancestors create advanced civilizations, say, three million years ago? In my opinion, they didn't do so because they simply weren't smart enough. I think our current level of intelligence is probably just barely sufficient to develop modern societies. If collectively our intelligence were to drop appreciably, human societies would crumble.

As a thought exercise, consider a human society that consists *entirely* of people with intelligence well below average. Could this society develop or maintain a level of civilization even close to what we have today? I'm very confident that the answer is no. There would be no competent teachers or leaders in such a society.

What about a society that consists entirely of people with average intelligence? I'm also confident that such a society would not be able to develop or maintain a high level of civilization. The absence of above average intelligence professionals would be an enormous, probably insurmountable, obstacle.

What about a society that consists entirely of people with above average, but not exceptional, intelligence? Here, I'm uncertain. Maybe the absence of geniuses, people like Newton and Einstein, with intelligence far above average, would impede development and maintenance of that society.

Real human societies, of course, do not consist entirely of people with one level of intelligence, but rather people with a wide range of intelligence. I don't think anyone knows what mix of intelligences is sufficient to maintain our societies, but very likely, societies with a higher average intelligence and with a larger fraction of highly intelligent individuals have a better opportunity to thrive.

For millennia, scholars have attempted to understand why some human societies are more successful than others (for just a tiny sampling of such attempts, see references 16–23). Although these scholars have identified many of the factors involved, I am certain that no one understands the matter completely. There are too many factors, and they interact in too many complex ways. I am convinced, however, that one important factor, one that is often neglected, is the distribution of intelligence in the society (24–26).

Many human traits, such as adult height or blood pressure, follow normal distributions (Figure 6.1). One property of normal distributions is that a relatively small change in the mean (the highest part of the curve) leads to relatively large changes at the extremes (Figure 12.1). This is just mathematics, as reliable as 2 + 2 = 4. It is not known whether intelligence closely follows a normal distribution at the extremes, but if it does, then small changes in the average intelligence in a society could result in a big difference in the number of exceptionally intelligent people, which could have a significant impact on that society.

I worked with top-level research scientists all my adult life. Top-level scientists are in many ways like all other people. They are tall and short, slender and stout, handsome and plain. They are also kind and vicious, reckless and cautious, extroverted and introverted. But top-level research scientists differ from most other groups in that they are, without exception, highly intelligent. Harness these individuals to specific goals, and they will accomplish amazing things.

I was fortunate in my life to have observed one such project firsthand. The sequencing of the first human genome in what was called the Human Genome Project was the organized international effort to map and sequence the first human genome. It lasted roughly 15 years, from about 1986 to 2001. At its height, the Genome Project employed no more than a few thousand people, yet the accomplishments of this group of scientists were spectacular. The project went from what seemed to be an outrageous goal in 1986 to a good draft sequence by

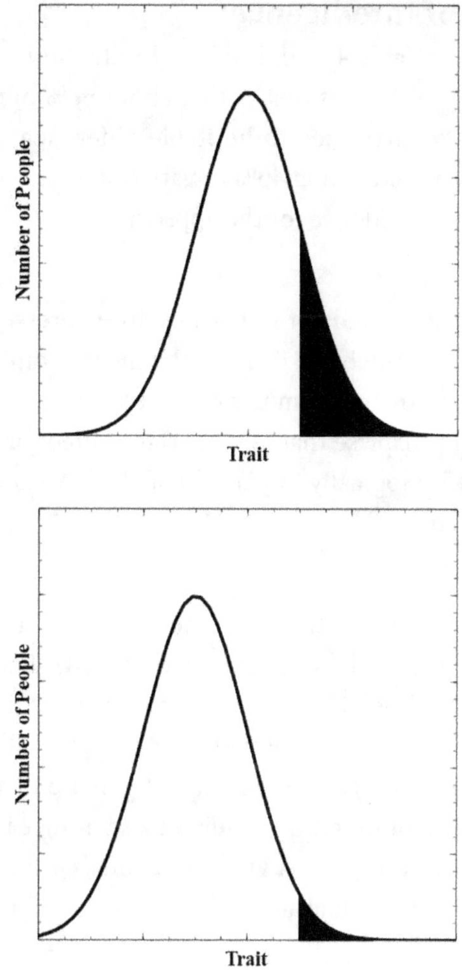

Figure 12.1 Changes in the averages of normal distributions
Both plots have exactly the same width, but the upper plot has a higher average than the lower plot. The sizes of the shaded areas represent the total number of people who have values of the trait above the level at the start of the shading. This level is the same in both plots.

2001. The project was completed ahead of schedule and under budget. I personally knew most of the project leaders and attest that they were exceptionally intelligent. I am certain that the project could not have been successfully completed with individuals of low or average intelligence. The challenge was too great.

Heritability of Intelligence

Back in Chapter 4 (Table 4.1), I listed the heritability of intelligence to be approximately 60%. This means that about 60% of the variation in intelligence can be attributed to heritable differences in our DNA. It also means that the remaining 40% of variation is due to nonheritable factors. What is the evidence for the approximately 60% heritability?

Biological Plausibility

It is known that thousands of our genes are expressed at significant levels in our brains. The brain is likely the most complex organ in our bodies. It is also known that mutation creates variability in all genes. Therefore, it is no surprise that DNA variants affect brain function. In fact, it would be biologically implausible for DNA variants to have no effect on intelligence.

Experiments with Lab Animals

Scientists have attempted to study intelligence in a number of laboratory animals, especially rats and mice (27–28). Lab rats and mice can be either inbred or outbred. Inbred strains have undergone many generations of brother-sister matings (an exceptionally severe population bottleneck) such that essentially all genetic diversity has been eliminated. Except for mutations, individuals within each inbred strain are identical in DNA sequence. Outbred animals are genetically more like humans. Each individual animal has an appreciably different DNA sequence.

A number of researchers attempted to breed rats with high and low intelligence (28–33). These researchers started with a group of outbred animals. They then ran the individuals through a test for "intelligence," usually some form of maze. In each generation, they took the highest-scoring individuals and bred them together, and took the lowest scoring individuals and bred them together. This process was continued over multiple generations. Invariably, these experiments resulted in "maze-bright" and "maze-dull" strains of rats that differed significantly from each other in performance on the tests and that bred true. Results from one such experiment are shown in Figure 12.2. Since all the rats in these experiments were raised under exactly the same conditions, there was no environmental variation. The bright and dull strains therefore must have differed in their germline DNA sequences.

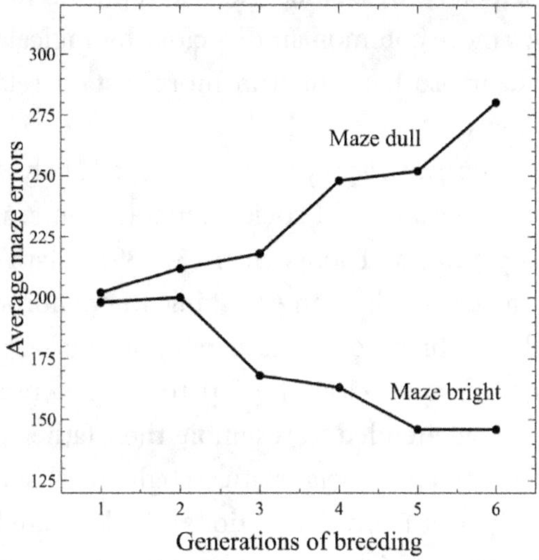

Figure 12.2 Selective breeding of maze-bright and maze-dull lab rats
(adapted from reference 32)

Typically, with lab mice, an alternative approach was used. Different inbred strains of mice were run through mazes and other tests of "intelligence" (34–35). The strains differed significantly in their performance on the tests. Again, since all the mice were raised under identical conditions, the differences in maze performance must be due to differences in DNA sequences among the inbred strains.

Certainly, human intelligence and maze running in lab animals are not equivalent. Nevertheless, the experiments with lab animals provide evidence for the heritability of intelligence.

Family Studies

If intelligence has substantial heritability, then children should resemble their parents in intelligence. Francis Galton was probably the first to attempt to show (in about 1865) that exceptional life achievement runs in families (36). Galton subjectively identified men in fields such as law, warfare, science, and the arts who rose to especially high positions and/or were exceptionally accomplished. He then examined the life achievements of male biological relatives of these illustrious men. He found that eminence (his term) in the relatives was much more common than expected by chance. He also reported that eminence in the relatives was more common among close biological relatives, such as sons, fathers, or brothers, than in more distant relatives, such as cousins or nephews.

Much more recently, Gregory Clark published analyses of father-to-son (patrilineal) inheritance of social status (income, wealth, education level, occupation, and longevity) (37–39). Clark's analyses are largely based on *rare* family names—think Buddenbrock, not Baker. Clark reported that high social status stubbornly persisted through patrilineal lines, in some cases for up to 10 generations or more. Thus, sons on average tended to resemble their fathers, grandfathers, great-grandfathers, etc. in social status. High social status did decay, but only gradually over many generations. Clark posited that the slow decay was due to substantial inheritance of talent, as well as marriages between high-status men and high-status women. Clark's reasons for excluding environment as the primary cause for the persistence in social status were that government social spending, such as subsidized education, has not markedly altered social mobility, and that windfall wealth, such as winning a lottery, does not translate to high social status in succeeding generations.

More concrete evidence that relatives resemble each other in intelligence comes from the degree of similarity in intelligence test scores for various pairs of relatives (40–41). The numbers shown in Table 12.1 were taken from a comprehensive review of over 100 different studies (40, Figure 1). Correlation is the resemblance in scores among the pairs of relatives. The higher the correlation, the closer the scores for pairs of individuals within the groups. The results shown in Table 12.1 are what would be expected if intelligence has substantial heritability—correlations drop as the fraction of shared DNA decreases.

However, as described in Chapter 4, similarity in intelligence among relatives could be due to shared environment, rather than shared DNA. Siblings, for example, generally share a more similar environment than cousins. And because they are the same age, twins likely share a more similar environment than other sibling pairs. Scientists have therefore turned to twin and adoption studies to disentangle heredity and environment.

Table 12.1
Correlation of intelligence test scores among various pairs of individuals*

Relationship	Correlation**	Fraction of DNA Shared
Identical twins	0.86	100%
Fraternal twins	0.60	50%
Midparent-children***	0.50	50%
Full siblings	0.47	50%
First cousins	0.15	12%
Unrelated individuals	0.00	0%

*Results for twins, midparent-children, and full sibs are mostly for individuals reared together. Results for first cousins and unrelated individuals are mostly for individuals reared apart.

**Weighted averages of various studies.

***Midparent-children is defined as a comparison of the average scores of the parents to the scores of their children.

Twin Studies

Since identical twins share 100% of their DNA, while fraternal twins share only 50%, it's expected that for any trait with substantial heritability, identical twins will be closer to each other than fraternal twins. Dozens of studies involving thousands of twin pairs from several countries have compared measures of intelligence between identical and fraternal twins (40–43). All of these studies have reported greater similarity in intelligence between identical than between fraternal twins. The numbers shown in Table 12.1 are representative of these studies.

Adoption Studies

Another standard approach for separating the influence of genes and environment is the study of adopted children. If heritability of intelligence is low, and environment is most or all, then the intelligence of adopted children should more closely match that of their adoptive parents (with whom they grew up) than their biological parents (with whom they did not grow up). However, if shared DNA is the predominant factor, then adopted children should more closely match their biological parents. Many studies carried out over decades have shown that adopted children more closely resemble their biological parents in intelligence, especially as the children get older (4, 44–47).

An approach that combines twin and adoption studies is the study of twins raised apart (48–50). Since twins are relatively rare, and twins raised apart are exceptionally rare, these studies are hampered by small numbers. Nevertheless, the results of such studies are illuminating. Reunited identical twins, even after decades of living apart, were found to be very similar in intelligence test scores. Correlations between 162 identical twin pairs reared apart from four studies completed over five decades are shown in Table 12.2. The correlations are all high, and are close to those reported for identical twins raised together (Table 12.1).

Table 12.2
Correlations in intelligence test scores for identical twins raised apart

Study	Numbers of Twin Pairs	Correlations
1	19	0.68
2	12	0.64
3	38	0.74
4	48	0.69
5	45	0.78
Totals	162	0.73

Taken from reference 50, p. 189.

Genes Involved in Intellectual Disability

In genetic terms, intelligence is clearly a complex trait. Variants in thousands of different genes have been shown to affect intelligence. However, like many other genetically complex traits, intelligence has single-gene subforms (Chapter 4). These subforms occur at the lowest end of the intelligence range, or what is typically called intellectual disability.

Intellectual disability (abbreviated ID and formerly called mental retardation) is defined as having intelligence test scores that are within the lowest 2.5% of scores (51–53). ID is therefore fairly common. Before children reach the age when intelligence tests become reliable, ID is often identified by developmental delay, which means the child is not hitting developmental milestones like walking and talking at the same age as most other children. While ID is not exactly the same as developmental delay, there is substantial overlap.

As covered in Chapter 4, each single-gene disorder can be described by one or more clinical features (see, for example, Table 4.5). The largest compilation of single-gene disorders lists ID as a clinical feature for over 1,000 single-gene disorders, and developmental delay as a clinical feature for over 3,000 single-gene disorders (54). Therefore, brain function is adversely affected by deleterious variants in a minimum of 5%, and probably more than 15%, of all genes. These variants have a mix of dominant, X-linked, and recessive inheritance (55–63).

Children with ID are often born to parents with normal or high intelligence. This can happen due to recessive inheritance or environmental factors, but is more often due to mutation. Roughly half of all ID cases are attributed to new, heterozygous variants that arise through mutation (62, 64–66) (see Case 2 in Chapter 3). All types of mutations, including nucleotide substitutions, can cause ID. This is an important and weighty observation: a change in a single nucleotide out of six billion can spell the difference between a person with normal or high intelligence and a person with ID.

Although ID is often caused by variants in single genes, variation in intelligence within the normal range shows different genetics (67). Thousands of variants are involved in normal intelligence, each with relatively small effect. Most of these variants were identified by analyzing the DNA of large groups of individuals, ranging from tens of thousands to millions, for which estimates of intelligence were available (68–72). Because intelligence tests are rarely, if ever, administered to such large groups these days, educational attainment, which is relatively easy to obtain, was often substituted as a crude measure of intelligence. Variants involved in intelligence were identified by differences in frequency among the groups with different levels of educational attainment. For example, a variant might have the highest frequency in the group of subjects who did not complete high school education, a lower frequency in those who attended college, and the lowest frequency in those with postgraduate degrees. The differences in variant frequencies were typically small, and could be identified only through studying large numbers of individuals.

Using the variants that influence educational attainment, indices have been constructed that explain about 10% of the variation in educational attainment (45, 72). Although average results from these indices appear impressive (Figure 12.3), individual results are far less so (Figure 12.4). In other words, although current polygenic indices for educational attainment *on average* distinguish those with high intelligence from those with low intelligence, their predictive power for any one person is quite limited. Many people with high indices have low educational attainment, and many with low indices have high educational attainment.

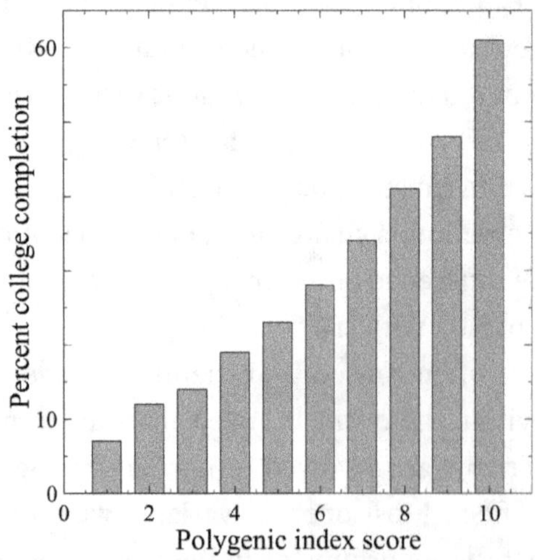

Figure 12.3 Predictive power of a current polygenic index for educational attainment: average results
(adapted from reference 72)

Nearly all the DNA variants from the large group studies that affect intelligence have relatively high population frequencies. This contrasts with the many single-gene variants that cause ID and have very low population frequencies. Recently, researchers have also begun to identify variants with intermediate population frequencies that also affect intelligence (73–77). Combining the effects of rare, intermediate, and frequent variants enhances the predictive power of polygenic indices (75–76). It is likely that the predictive power of the indices for intelligence will continue to improve in coming years.

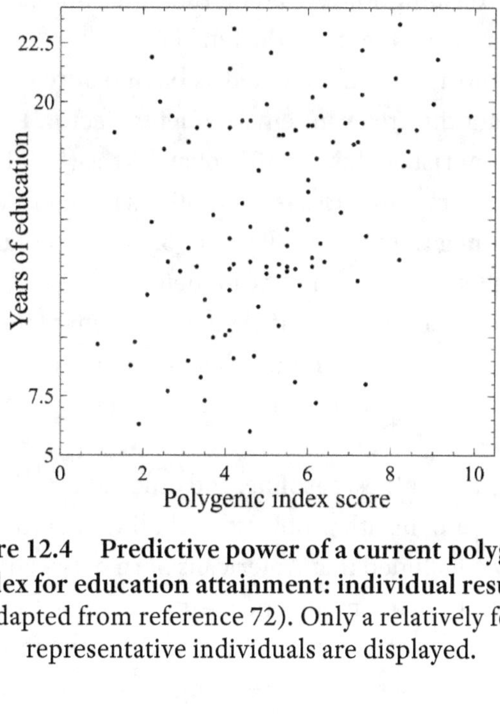

Figure 12.4 Predictive power of a current polygenic index for education attainment: individual results (adapted from reference 72). Only a relatively few representative individuals are displayed.

Nonheritable Factors

Because this book is about genetics, I emphasized the heritability of intelligence. But it's important to keep in mind that roughly 40% of variation in intelligence is due to nonheritable factors.

As presented in Chapter 4, nonheritable factors include germline mutations, somatic mutations, chance events, and environment. The strong impact of some germline mutations on intelligence is described above. Although less is known about the impact of somatic mutations, there are early indications that they also play a role (78–79). It's also likely that random events in fetuses and children, such as X chromosome inactivation in females, affect development of the brain. Finally, a number of environmental factors have been linked to variation in intelligence. I'll cover just a few of these.

Excessive consumption of alcohol by a pregnant woman can result in a child with fetal alcohol syndrome (80–81). Fetal alcohol syndrome is characterized by a number of health problems, including ID. Even when alcohol consumption by the mother does not cause frank ID, it may still impair development of the fetal brain.

Premature birth, typically defined as birth before the 37th week of pregnancy, is another relevant environmental factor. Premature birth is common, comprising close to 10% of all births (82). Like fetal alcohol syndrome, premature birth is associated with a host of problems in the child, including reduced intelligence (83–84). The earlier the birth, the greater the likelihood of significant deficits.

Malnutrition in children can affect development of their brains (85). Diets deficient in a number of individual nutrients, like iodine, are still common in developing nations. These deficiencies can cause cognitive deficits (86–87).

Lead from paints, pipes, gasoline, and other sources has been shown to cause a number of health problems, including cognitive deficits (88–89). One study concluded that Americans born between 1951 and 1980 lost an average of 1% on intelligence test scores due to lead poisoning (88). People living in low-income areas of US cities have generally received higher lead doses than those living in affluent areas.

Good evidence also exists that upbringing has an effect on intelligence. Rats raised in a "enriched" environment that included ramps, mirrors, balls, and tunnels performed significantly better on maze tests than genetically identical rats raised in "restrictive" environments without any of these accoutrements (90). Human children raised in extremely deprived environments have cognitive deficits (91–92). Children adopted into high-status homes demonstrate significant improvements in intelligence test scores compared to biological siblings raised with their biological parents (4, 93–96).

Average raw intelligence test scores have increased substantially throughout much of the 20th century in many countries (4, 97–99). This improvement in scores has come to be known as the Flynn effect, after James Flynn, who was one of the first to report this phenomenon. Since changes to the gene pool are unlikely to explain the Flynn effect (Chapter 14), changes in environment are the likely explanation. General improvements in diet and health are probably at least partly responsible; average adult height also increased significantly during the same period. Flynn himself concluded that the primary cause was improvements in test-taking skills (100). A number of researchers have recently reported that the Flynn effect has plateaued or even reversed in developed nations in the late 20th and early 21st centuries (101–105).

Conclusions

Virtually no one disagrees that tall parents tend to have tall children, and short parents tend to have short children. People also have little trouble agreeing that genetically complex health problems, such as cancer or heart disease, often run in families. Yet, when it comes to intelligence, people often reflexively reject even the possibility of heritability. This is illogical and wrong.

The lines of evidence in support of the substantial heritability of intelligence are summarized in Table 12.3. Individually, each of these lines provides compelling evidence for the substantial heritability of intelligence. Together, they comprise an overwhelming case.

Other scientists who have reviewed the genetics of intelligence have also concluded that intelligence has substantial heritability (4, 45, 106–108). The most thorough, balanced summary I have found was written by N. J. Mackintosh (4). Mackintosh carefully described the evidence, along with the limitations of the methods employed. I don't know of any reputable human geneticist who today disputes the substantial heritability of intelligence. The substantial heritability of intelligence is not hearsay, nor opinion, nor hypothesis, but *scientific fact*.

Table 12.3
Sources of evidence for the substantial heritability of intelligence

- Biological plausibility
- Experiments with lab animals
- Resemblance of family members
- Twin studies
- Adoption studies
- Single-gene subforms of intelligence (intellectual disability)
- Polygenic indices for intelligence
- Scientific consensus
- Evolution of human-level intelligence (next chapter)

As mentioned in Chapter 4, heritability is difficult to measure, and there is no perfect method for doing so. Different studies have obtained estimates for the heritability of intelligence that range, with few exceptions, between 40% and 80%. The 60% number I presented in this book is a conservative midpoint of these estimates.

But regardless of the true numeric value, the important point is that heritability of intelligence is substantial. For the purposes of this book, it matters little whether it is 40% or 80%. What is important is the fact that our DNA sequences strongly influence our intelligence, and that children usually resemble their parents in intelligence.

In contrast to geneticists' knowledge of variants that cause ID, next to nothing is known about the genetics of high intelligence. No variants have yet been identified to explain exceptional intelligence. I don't know of any cases in which a brilliant child was born to parents of limited intelligence as a result of mutation. My best guess is that exceptional intelligence involves having a fortuitous combination of relatively few deleterious variants that impair brain function and favorable nonheritable factors, but this is just my educated guess. We need to learn much more about the genetics of intelligence (Chapter 17).

Since about 2% of people have ID, and half of ID cases are caused by germline mutation, mutation is responsible for ID in roughly 1 out of every 100 newborns. This is not too surprising when we consider that an average of two mutations occur within each baby's genes, that many of these mutations are deleterious, and that probably 15% or more of our genes are involved in brain function. However, in addition to the mutations that cause frank ID, there are many other mutations that cause a smaller decrease in intelligence. Mutations have a wide range of impacts (Chapter 3). It is possible that as many as 1 in 20 or even 1 in 5 newborns has a significant decrease in innate intelligence due to mutation (109). Germline mutation relentlessly erodes the intelligence of human populations.

Summary

- Intelligence, broadly defined as the ability to solve problems, varies over a wide range among individuals.
- Higher intelligence is a significant factor in both individual and societal success. Human civilizations could not have arisen without our special intelligence.
- There are multiple compelling lines of evidence that intelligence has significant heritability. These include family studies, twin studies, adoption studies, and the genetics of intellectual disability (ID) (Table 12.3).
- Geneticists have now identified the specific DNA causes of many thousands of cases of ID. A change in a single nucleotide out of six billion can spell the difference between a person with normal intelligence and a person with ID.
- Although different studies have produced a range of estimates for the heritability of intelligence, it is a scientific *fact* that intelligence has substantial heritability (on the order of 60%).
- Nonheritable factors also strongly influence intelligence. These include mutations and environmental factors such as premature birth, lead poisoning and upbringing.
- The mutations that occur in every child in every generation (Table 3.3) cause frank ID in about 1 out of every 100 newborns, and somewhat reduced innate intelligence in an unknown larger fraction of newborns.
- Germline mutation relentlessly erodes the intelligence of human populations.

Chapter 13

EVOLUTION OF OUR SPECIAL INTELLIGENCE

This chapter focuses on the evolution of human special intelligence, with emphasis on the forces that shaped this evolution.

Human-Ape Common Ancestors

Today, humans dominate the globe. But this was not always the case. Roughly eight million years ago, our ancestors were a relatively small population of ape-like creatures (Chapter 5). These creatures were the common ancestors of chimpanzees, gorillas, and humans. First, the ancestors of gorillas split from the ancestors of chimps and humans, and then about seven million years ago, the chimp and human lines split (Figure 5.2).

Very little is known about the common ancestors of humans and apes. Only a few fossils that could (but are not certain to) be from the common ancestors have been discovered (1–4). However, it is known that over the millions of years since the common ancestors, human anatomy changed considerably (Figure 5.4). The arms of our ancestors shortened and their legs lengthened. Their muzzles receded. Most importantly for this chapter, their brain volume increased about threefold (5–8). Artists working with paleontologists have attempted to create images of how our ancestors may have looked based on the fossil bones. Although they are only approximations, these images are still fascinating (9).

Chimps and gorillas are probably the most intelligent animals outside of humans. It's therefore likely that the common human-ape ancestors were relatively intelligent. But the intelligence of the common ancestors did not match our own. There is absolutely no evidence that they created tools, built structures, or manufactured items. Over the last seven to eight million years, our ancestors evolved the special intelligence that allowed us to dominate the world.

Available Evidence Relevant to Evolution of Human Intelligence

The major sources of evidence relevant to the evolution of human intelligence are listed in Table 13.1. Some evidence comes from the behaviors of living wild and captive chimpanzees and gorillas. However, these behaviors can be misleading. Chimps and gorillas are not identical in their behaviors. Also, the common human-ape ancestors were not humans nor chimps nor gorillas, but different creatures (2). So, at best, the behaviors of apes living today may provide vague clues.

Many thousands of ancient bones dating from as far back as eight million years have been discovered. However, most of these remains are fragments, rather than full skeletons, and as the specimens go further back in time, they become scarcer (10).

Nonbiological evidence includes the stone tools constructed by our ancestors that have been dated as far back as 3.3 million years (11–12). As time passed, these tools slowly became more sophisticated. There is also evidence of fire usage dating from about 400,000 years ago, perhaps earlier (13–14). Archaeological artifacts such as cave paintings, jewelry, sculptures, remains of buildings, and even musical instruments have been discovered going back about 50,000 years, with a few earlier specimens (15). And there are written historical records of human societies accumulated over the last several thousand years.

DNA has been extracted and sequenced from many human and human-like remains (Chapter 5). Over 100,000 ancient DNA sequences have now been generated, a few dating to over 50,000 years ago (16–17).

The results of modern anthropological studies on isolated hunter-gatherer populations who were still using Stone Age technology are also available. Most of these studies were carried out in the 20th century.

Table 13.1
Major sources of evidence about human evolution over the last several million years

Evidence	**Time Span Dating Back from Present (Years)**
Behaviors of extant wild chimps and gorillas	8 million
Fossil bones	8 million
Stone tools	3.3 million
Ancient DNA sequences	200,000
Archaeological artifacts	50,000
Historical records	10,000
Modern study of hunter-gatherer populations	200

The volume of evidence relevant to the evolution of human intelligence decreases as we go back in time. For a large part of the eight million years since the common human-chimp-gorilla ancestor, very little information is available. Most of this time is therefore dark to us.

What Is Known About the Evolution of Human Intelligence

Although a great deal is still unknown, three important facts have been established about the evolution of human special intelligence. First, evolution of our special intelligence primarily involved changes to our genomes. Undoubtedly, establishment of human civilization was in part dependent upon cultural and technological developments such as domestication of plants and animals, written language and smelting of iron. But these developments would not have been possible without changes in the genome. If learned culture and technology alone were responsible for our civilizations, then chimps and gorillas raised as humans should be able to function like humans. This experiment has been performed multiple times (see, for example, 18–19), but the result is always the same. Although the great apes are remarkable animals, chimps and gorillas raised as humans grow up to become adult chimps and gorillas, not adult humans. They can't write, they can't do mathematics, and they can't build computers.

Second, the new DNA variants that led to improved intelligence arose through mutation. This is an obvious point, but still important to keep in mind. By comparing genome sequences from gorillas, chimps, Neanderthals, and modern humans, researchers have tried, and continue to try, to determine which changes in our genomes have led to our special intelligence (8, 20–23). It turns out, however, that this is a very difficult problem. Despite the high degree of sequence similarity between the genomes of these four species (Table 5.2), the numbers of nucleotide differences are still daunting. The human and chimp genomes, for example, differ by more than 40 million nucleotides. Although researchers have identified some sequence differences that *may* be important, this approach still leaves us far from understanding the evolution of our special intelligence.

And third, over the long haul, the more intelligent of our ancestors reproduced at higher rates than our less intelligent ancestors (24). Gradually, over millions of years, mutation created new sequence variants that caused some of our distant ancestors to be more intelligent. Because individuals who carried these new variants left more descendants than those who did not, the population frequencies of the new variants increased. Eventually, many of the new variants reached 100% population frequency. This would not have been possible without the differences in reproduction.

In addition to these three facts, there is a fourth important point that, although not certain, is I think very likely. From the work that has been done to date comparing genome sequences among our closest relatives, it can be concluded with reasonable certainty that evolution of our special intelligence was not due to just a handful of variants in a few genes, each with relatively large effect. Rather, it was due to the gradual accumulation of many variants, each with relatively small effect (20–22, 25–26). Many of these variants are located outside of exons, within gene regulatory regions. A *gradual* increase in the intelligence of our ancestors is also consistent with the fossil (increasing brain size) and stone tool records.

Forces Involved in the Evolution of Human Intelligence

The process through which our special intelligence evolved was undoubtedly highly complex. No one, certainly not myself, currently has a good understanding of this process. I think that the best that can be done at this point is to describe some of the major forces that were involved. The forces that I have identified are listed in Table 13.2. The remainder of this chapter is devoted to a one-by-one description of these forces, along with some thoughts on the role they may have played in the evolution of our intelligence. There is *substantial* overlap among many of these forces.

Table 13.2
Primary forces involved in human evolution

Primary forces involved in human evolution
Differential Reproduction
Natural Selection
Genealogical Demography
Polygyny
Population Size
Migration
Infanticide
Assortive Mating
Population Crashes
Mutation
Chance

Differential Reproduction

As long as a population does not have appreciable immigration or emigration, a variant within that population can only increase or decrease in frequency if carriers of the variant reproduce at higher or lower rates than those without the variant. As previously mentioned, our ancestors who carried variants that increased intelligence *must* on average have reproduced at higher rates than those without such variants. Furthermore, because throughout evolutionary history mutation has constantly created variants that decreased intelligence, ancestors who carried such deleterious variants must have reproduced at lower rates than those without them.

Natural Selection

Natural selection means that those individuals (of any species) that are best adapted to their environment are the ones most likely to survive and reproduce (Chapter 3). There is considerable evidence that natural selection has shaped human genomes. Let's start with negative natural selection. Evidence for negative selection comes from the finding that DNA variants that disrupt the function of many genes are depleted in the current human gene pool (27–28). Shown in Table 13.3 are observed numbers divided by expected numbers (ratios) for selected exonic variants of three types: loss-of-function variants, which generally abolish gene function; missense variants, which lead to changes in amino acid sequences in proteins; and synonymous variants, which do not alter amino acid sequences (Chapter 3). Expected variant frequencies were determined by assuming random mutation and no natural selection. Loss-of-function variants are depleted more than missense variants, while synonymous variants show little or no depletion.

Comparing the genome sequences of different species provides additional evidence for negative selection. When DNA sequences of vertebrates are compared, for example, exon sequences are found to be much more conserved than intronic and intergenic sequences (29–30). This is because deleterious exonic variants generally have a stronger adverse effect on the health and abilities of individuals than deleterious variants located outside of exons, and are therefore more efficiently removed by negative selection. Furthermore, because of the redundancy in the genetic code (Figure 2.9), amino acid sequences of proteins are more highly conserved among different species than exonic DNA sequences (31). Changing the amino acid sequence of a protein often adversely affects the protein's function, but exchanging one redundant

DNA code for another does not alter the amino acid sequence or the function of the protein.

Table 13.3
Ratios of observed to expected variants in selected genes by variant type

Gene	Disorder	Loss-of-function	Missense	Synonymous
DDX3X	Intellectual disability	0.00	0.28	1.09
FBN1	Marfan syndrome	0.08	0.63	0.90
PTPN11	Noonan syndrome	0.08	0.51	0.83
CHD7	CHARGE syndrome	0.08	0.82	0.99
ARID1B	Intellectual disability	0.15	0.96	1.16
TSC1	Tuberous sclerosis	0.16	0.74	0.87
NF1	Neurofibromatosis	0.38	0.60	0.89
RYR1	Myopathy	0.59	0.87	0.98
G6PC1	Glycogen storage disease	0.66	0.85	0.88
BRCA1	Breast cancer	0.77	0.87	0.83

Data taken from the gnomAD database (https://gnomad.broadinstitute.org/).

Studies in lab animals called mutation accumulation experiments offer further evidence for negative selection. In these complex and difficult experiments, mating conditions are arranged so that nearly all mutations, except for those incompatible with life and reproduction, are allowed to accumulate over multiple generations (32–35). The lines of animals (worms, flies, and mice) in which the mutations accumulate are weaker and less fertile than control animals in which deleterious variants were removed by negative selection.

Positive natural selection leads to increases in the frequencies of advantageous variants. Several examples of this phenomenon in humans have been discovered (36–39), including the variants that cause lactase persistence and resistance to malaria (Chapter 3). I do not know of any confirmed reports of positive selection for variants that increase intelligence. If, as I think likely, there are many such variants, each with a small effect, then signals of positive selection for intelligence may be too weak to detect using current methods. However, it's easy to imagine how natural selection could have worked in our ancestors to gradually increase intelligence. Individuals who were more intelligent could more efficiently obtain food and mates, and better plan attacks and defenses against neighboring groups. And perhaps most importantly, more intelligent parents would be more likely to raise children who survived to reproductive age than less intelligent parents.

Natural selection is not a perfectly efficient process. Often, especially in the past, a child who was especially healthy and bright died because of a chance event such as a drought or a storm. And sometimes, a less healthy and talented child happened by chance to be born into a group that was relatively prosperous and peaceful, and therefore survived and reproduced. But over the long term, those who were healthier and more talented produced more descendants and contributed more to the gene pools of subsequent generations.

Genealogical Demography

I define genealogical demography as a branch of demography concerned with the numbers and characteristics of ancestors and descendants of individuals. In the next chapter, I focus on descendants of people who are currently living; in this chapter, I focus on ancestors. For adults of reproductive age living at any time in the past, geneticists would like to know the contributions of those individuals to the current gene pool. Genealogical demography exerts a strong influence on the gene pool.

Consider two famous people from history who lived at about the same time: President Abraham Lincoln in the US (1809–1865) and Queen Victoria in England (1819–1901). Lincoln and his wife, Mary, had 4 sons, but only 1 of the 4 (Robert) survived to adulthood. Robert had 3 children of his own, but Robert's children either had no children, or their children (President Lincoln's great-grandchildren) had no children (Figure 13.1). Therefore, none of Abraham's and Mary's DNA is present in today's gene pool. In contrast, Victoria and her husband, Albert, had 9 children who lived to adulthood and 42 grandchildren (Figure 13.2). Today, Victoria and Albert have roughly 1,000 living descendants. They therefore made an appreciable contribution to today's gene pool.

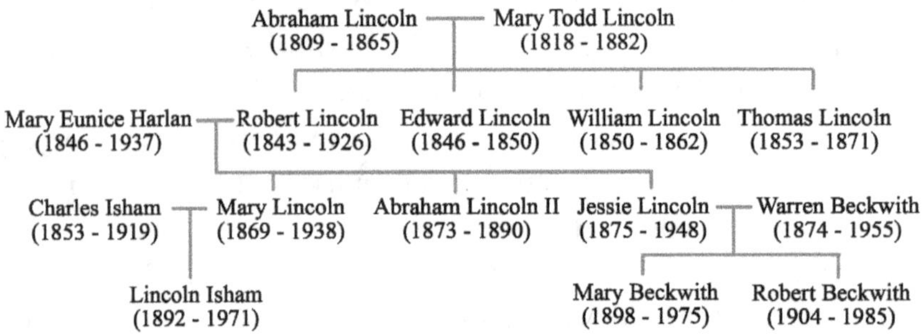

Figure 13.1 Descendants of Abraham and Mary Lincoln

Figure 13.2 Queen Victoria's family
Shown are Victoria, her husband, Albert, their (adult) children, and many of their grandchildren.

Although only limited information is available about the number of descendants of people who lived in the past, two firm conclusions can still be made. First, many adults who lived in the past made no contributions at all to the current gene pool (40–43). Throughout all of human history, there have been many individuals who were childless, due, for example, to infertility or the inability to find a reproductive partner. Also, in the past, a large fraction of the children who were born did not survive to reproductive age. Even for those who had children who reached reproductive age, there were many, like Abraham and Mary Lincoln, whose line of descent winked out after just one or a few generations.

A study of four generations of people living in northern Sweden from 1885 to 2007 found that 48% of the people in the first generation (age 15 and above) did not have any living great-grandchildren (44). Going back 8–10 generations, there is evidence that the fraction of people who did not contribute any of their DNA to the present was more than 50% (45–46). And another study concluded that of the people living at the dawn of modern humans, more than 90% did not contribute to the current gene pool (47).

The second firm conclusion is that for those from the past who did contribute to the current gene pool, their contributions varied widely. Some contributed only a small amount and others a great deal. This conclusion came from the study of established genealogies (43–46, 48–49) and also from what are known as founder variants.

Founder variants are variants, usually causative for recessive disease, that are found predominantly in a single population. Frequencies of founder variants are *relatively* high in those populations. Populations that harbor founder variants were, at least in the past, reproductively isolated. Members of these populations reproduced almost exclusively among themselves, as opposed to reproducing with people outside of the populations. Reasons for reproductive isolation included geographic separation and cultural and religious practices. A small sampling of the many known founder variants is presented in Table 13.4.

Every mutation arises on a particular chromosomal background of nearby variants. When geneticists examine the backgrounds of founder

Table 13.4
Examples of founder variants

Population	Gene	Disease	Variant	Approximate Current Number of People with the Variant
Europeans	*CFTR*	Cystic fibrosis	ΔF508	10 million
Ashkenazi Jews	*HEXA*	Tay-Sachs	c.1275_1278dup	260,000
East Asians	*PAH*	PKU	p.Arg413Pro	200,000
Amish	*APOB*	Hypercholesterolemia	p.Arg3527Gln	50,000
French Canadians	*SACS*	Ataxia	c.8844del	30,000
Finns	*SLC26A2*	Diastrophic dysplasia	c.-26+2T>C	25,000

References: 50–57.

variants, they find that, with rare exceptions, each person who carries a specific founder variant has the same chromosomal background. This means that nearly all people living today with these variants are descended from the *single* person in whom the mutation originally occurred. As an example, almost all of the approximately 10 million people living today who carry the ΔF508 variant in the *CFTR* gene are descended from an individual who is thought to have lived about 50,000 years ago (50).

What were the people like who lived in the past and who made especially large contributions to the current gene pool? Unfortunately, almost nothing is known about them. But there is evidence that, on average, those who had large numbers of surviving children were more likely to make any contribution (45) and to also make greater contributions (44). This makes intuitive sense. The line of descent from Victoria and Albert, who had nine surviving children, continues to this day, while the line from Abraham and Mary, who had only one surviving child, ended. There is also evidence, although the effect is relatively weak, that children from large families also tended to have large families (45, 54, 58–59). Those of our ancestors whose descendants favored large families over multiple generations contributed more than most others to the current gene pool.

In summary, the contributions of people who lived in the past to the current gene pool range from zero to exceptionally large. However, we should also keep in mind that the genome of any single person living today is a composite of DNA from many different ancestors (60). Susumu Ohno calculated that about 4,000 ancestors who lived 20 generations ago contributed to the genome of each person alive today (61). According to Ohno, we are all descendants of local kings as well as "murderers, thieves, embezzlers, prostitutes, and all other social misfits of the times."

Polygyny

Polygyny is defined as males reproducing with multiple female partners. Polygyny contrasts with polyandry, in which females reproduce with multiple male partners, and monogamy, in which males and females reproduce with only a single partner. Remember from Chapter

11 that males have virtually unlimited reproductive potential, but females can only have limited numbers of children. Polygyny may have been a significant force in the evolution of our special intelligence.

Polygyny is present in both of the African great apes. Gorillas have a harem reproductive pattern (62–63). A dominant male will reproduce together with a group of about five females whom he aggressively controls. When a new male defeats the previous dominant male, he will gain control of the females and may even kill the young offspring of the previous dominant male. Although chimps do not have a harem structure, males live within a clear pecking order. Dominant males have more sex with females and almost certainly sire more offspring (62, 64). Although no direct evidence is available, it therefore seems likely that our distant ancestors practiced polygyny.

By comparing the evolution of the Y chromosome, which is inherited solely from fathers to sons, to the evolution of mitochondrial DNA, which is inherited over multiple generations solely from mothers to daughters (Chapter 2), geneticists have concluded that the number of males who contributed to the current gene pool were fewer than the number of contributing females (65–66). This result indicates that polygyny was likely practiced in anatomically modern humans.

In the past, and still today, deceased family members are sometimes buried together at the same site. Sequencing of DNA from bones found in ancient cemeteries has recently allowed family trees to be constructed. In several cases, these ancient family trees, ranging from about 5,700 to 1,300 years ago, demonstrated polygyny (67–70).

From studying modern hunter-gatherer populations, anthropologists have concluded that polygyny was common in preagricultural human societies (63, 71–72). Dominant males often had multiple wives, while non-dominant males often had one or no wives. From the study of some pre-metal populations in the Amazon, James Neel reported that there was large variation in the number of grandchildren descended from different adult males (72). Just considering those men who had grandchildren, the number of grandchildren per man ranged from 1 to 62. Furthermore, about half of the grandchildren were descended from only 13% of the men.

From historical records, it is known that polygyny was also common in agricultural societies (73). The transition from a hunter-gatherer to an agricultural lifestyle may even have been accompanied by an increase in polygyny (66, 73). It was not unusual for kings, emperors, etc. to have harems of hundreds or even over 1,000 females and to father hundreds of children. As just one of many examples, Moulay Ismail Ibn Sharif of Morocco (1645–1727) is reported to have had about 700 wives and concubines and to have fathered over 1,000 children (74–75). Although in Europe, the Catholic Church was effective in gradually eliminating polygyny, polygyny was frequently practiced in many other parts of the world until relatively recent times. In Japan and China, polygamy did not become illegal until 1880 and 1949, respectively.

From the study of Y chromosome DNA in men living today, geneticists have concluded that some men who lived in the past made spectacular contributions to the current gene pool. Probably the best-known example is Genghis Khan (1162–1227), the founder of the Mongol Empire. Genghis Khan had over 500 wives and concubines, many of whom were queens or princesses who were either taken captive or gifted to him. He is thought to have fathered hundreds of children. A study of the Y chromosomes of men living today in Asia indicated that 8% of the men share the same basic Y chromosome (76). This is roughly 20 million men, or 1 out of every 200 men on the planet. The authors of the study speculated that this common Y chromosome was directly inherited from Genghis Khan and/or his close male relatives. The geography, genetics, and history all fit this speculation. A similar study in Ireland identified a Y chromosome shared by about 300,000 men that is thought to have been descended from the medieval Uí Néill dynasty (77).

From past examples of polygyny, it is known that it was the leaders of groups who tended to have the most wives and who fathered the most children. Male leaders often achieved their positions through intense competition. Although I'm sure there were exceptions, male leaders were usually among the healthiest and most intelligent men (78–79). As Neel put it: "Dummies don't become headmen" (80). Because intelligent male leaders generally had the most wives and fathered the most children, polygyny probably played a major role in the evolution and maintenance of our special intelligence.

Finally, although numbers of offspring for women did not vary as much for men, it should not be ignored that women living in the past also differed in their numbers of children and contributions to the gene pool (81). More intelligent mothers were undoubtedly more successful, on average, in achieving survival of their children, and perhaps also in partnering with more intelligent men.

Population Size

Throughout at least much of human evolutionary history, people lived in relatively small groups of no more than tens to a few thousand individuals. Both chimps and gorillas live in small groups (62–64). Most hunter-gatherers who have been studied in modern times lived in small groups (82). Modern humans experienced population bottlenecks when they migrated out of Africa (Chapter 5). And after the development of agriculture, when by far the most common human occupation was farming, many people lived in small, rural communities.

When population sizes were small and travel over long distances was difficult, out of necessity people reproduced together with their biological relatives. This doesn't mean close relatives. Taboos on incest (83) (Chapter 15) and the practice of men acquiring their wives from other groups often tempered mating among close relatives, but small populations still meant that matings between, for example, second cousins were relatively common. Geneticists call such matings inbreeding. Ancient DNA studies have confirmed that inbreeding was common in our past (84–86).

From a genetic viewpoint, there are at least four primary outcomes of inbreeding. First, the frequency of recessive disease is higher in inbred compared to outbred populations (Figure 15.1). Second, as described in Chapter 3 (Table 3.4), inbreeding tends to reduce the number of rare variants in a population and overall genetic diversity (87–88). Third, inbreeding increases the rate of change of variant allele frequencies. Rare variants can rise in frequency relatively rapidly. Fourth, and most importantly for this discussion, studies in other organisms (89–90), as well as in humans (91–98), have shown that inbreeding generally reduces health and abilities. Although the effect is modest, intelligence is one of many human traits that has been shown to be adversely affected by inbreeding (92–94, 96–98).

Migration

There is abundant evidence that even in the absence of modern methods of transportation, some of our ancestors migrated over long distances. For instance, prior to the appearance of modern humans, and starting about two million years ago, *Homo erectus*—one of our ancestors (or close relatives)—migrated from Africa into much of Eurasia (99–100). Beginning about 60,000 years ago, modern humans migrated out of Africa to colonize the rest of the world (Chapter 5). Study of both ancient and extant DNA has shown that migrations of later, but still prehistoric, humans were also common (101). As just a few of many prehistoric examples, Europe has seen waves of migration transform the genomes of its populations (101–105). Bantu-speaking people migrated from western to eastern and southern Africa (106), and people from mainland East Asia migrated to the islands of Japan (107). There are also many examples of migration in historic times. Consider the migration of Europeans to the Americas and Australia, and the forced movement of African slaves to the Americas and other parts of the world.

Migration has always been accompanied by reproduction between individuals from the immigrating and native groups (101–105). In some cases, this intergroup reproduction was minimized by the immigrating group displacing the resident group and/or maintaining strictures on reproduction. But in other cases, there was extensive reproduction between the groups, and in all cases, there was at least some genetic mixing.

Migration, through reproduction among people who were previously genetically isolated, tends to counteract the effects of inbreeding. It increases the numbers of variants and the combinations of variants upon which evolutionary forces may act. Consider two variants, 1 and 2, each of which slightly increases intelligence. Variant 1 is unique to population A, and Variant 2 unique to population B. As long as populations A and B remain apart, then Variants 1 and 2 won't be found together in individuals, but when populations A and B mix reproductively, some of the children will inherit both Variant 1 and Variant 2, and may be more intelligent than either of their parents.

Infanticide

Infanticide is the killing of infants, usually by their parents, either directly or through abandonment (often called exposure). Although certainly an unpleasant subject, infanticide was practiced in the past by many, perhaps nearly all, human societies (72, 74, 108–110). Depending on the society and the circumstances, up to about 50% of newborns were killed.

The three main reasons for infanticide appear to have been shortage of food, illegitimacy, and preference for male children. It's difficult for us today to understand, but for most of the existence of modern humans, famine was a frequent companion. Mothers in hunter-gatherer societies often nursed their children for three to five years. When, for instance, a woman who was nursing a healthy two-year-old gave birth to another child, there often simply wasn't enough milk for two children. It was then a very hard, but pragmatic, decision to continue to feed the healthy child, rather than risk the lives of both children. In many human societies in the past, it was a serious crime for women to have extramarital sex; the penalty was often death. Therefore, many women who became pregnant outside of marriage hid their pregnancies and killed their babies as soon as they were delivered. Finally, especially when resources were limited, many human societies favored male children. The preferential killing of female babies usually resulted in more men than women of reproductive age. This meant more competition among males for female partners. Also, where women did have a choice, they could be more selective in choosing partners.

In societies where infanticide was common, even healthy, strong babies had limited odds of surviving to reproductive age. Babies that were physically abnormal had even less chance. Several investigators have reported that infants with congenital abnormalities were preferentially killed (72, 78, 108).

Assortive Mating

Assortive mating means that people often choose reproductive partners who are similar to themselves in various traits. Assortive mating has been demonstrated for traits such as age, height, weight, religious beliefs, political views, and intelligence (111–112). Intelligence is one of the traits that shows the strongest assortive mating. Because of the significant heritability of intelligence, assortive mating guarantees that human societies will continue to have some individuals of high intelligence in future generations. While assortive mating does not alter the average trait value for a population, it does broaden the spread of values (113–114) (Figure 13.3).

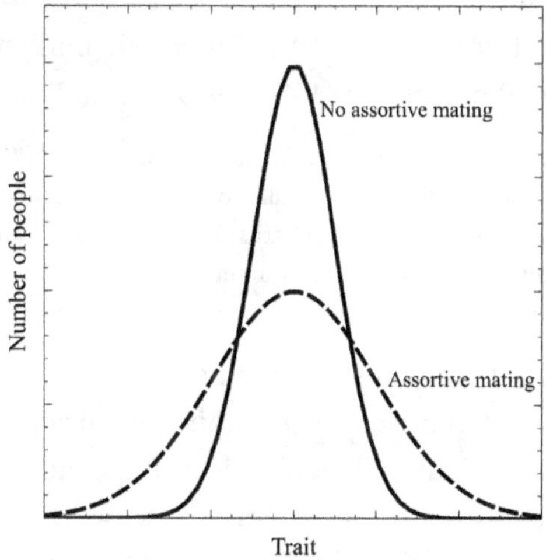

Figure 13.3 Assortive mating
The solid curve represents the population distribution of a trait in the absence of assortive mating, and the dashed curve represents the distribution in the presence of assortive mating.

Population Crashes

I define a population crash as a relatively rapid drop in the size of a human population. Although population crashes are rare today, they were frequent in the past. *Many* such crashes are known from recorded human history, and it is likely that they occurred frequently throughout all of human history (115–118). Just a few relatively recent examples are listed in Table 13.5.

Table 13.5
Selected relatively recent population crashes

Event and Location	Dates	Number of Deaths	Fraction of Population
Black Death in Europe	1347–1351	30 million	50%
Epidemics in Native Americans	~1500–1800	48 million	90%
Potato famine in Ireland	1846–1852	1 million	15%
Great Leap Forward famine in China	1959–1961	20 million	3%
Famine in North Korea	1995–2000	1 million	4%
World War II in Europe	1939–1945	42 million	10%
Genocide in Cambodia	1975–1979	2 million	25%

References: Black Death (117, 119–120); Native Americans (121); famines (118); WWII (116); Cambodia (122). Although all the events listed in the table were clearly population crashes, the numbers of deaths are only rough estimates and carry significant uncertainty. The Native American numbers are particularly uncertain.

Causes of population crashes include famines, wars, and epidemics. After the development of agriculture but prior to the dramatic improvements in agricultural yields beginning in about the year 1800, famine was probably the most frequent cause of crashes. This concept was most famously described by Thomas Malthus in 1798 (123). Malthus claimed that people were capable of reproducing much faster than food supplies could increase, and this led to boom-and-bust cycles. In good times, when food was plentiful, populations grew rapidly, but when drought, floods, or other factors reduced the food supply, people starved to death or curtailed population by other means, such as fewer and later marriages and infanticide. In this way, over the long term, population sizes didn't change much (Figure 14.1).

Probably the greatest effect of population crashes on the gene pool is the differential survival of people through the crashes. If people who possessed or were missing a particular variant preferentially survived a crash, then the population frequency of that variant could change appreciably and rapidly. Ancient DNA evidence was recently published indicating that those in Europe who survived the Black Death emerged with shifted frequencies for variants in some immune system genes (120).

Evidence is accumulating that healthier and wealthier people preferentially survived crashes (118–119, 124–125). This is especially well-documented for the 2020–2022 COVID-19 pandemic (124–125), during which roughly 20 million people died worldwide. At least in developed nations, those with lower levels of education and income died at higher rates. Higher-income individuals quarantined themselves by vacating cities and working from home. Many lower-income individuals had no choice but to continue working in crowded workplaces and therefore were more frequently infected.

A similar scenario likely played out during earlier crashes (118–119, 124). During famines, the wealthy could continue to buy food even when supplies were limited. They could also sometimes travel away from battlefields and disease-ridden cities. Wealth is only a crude proxy for intelligence, but it is still known that, on average, wealthy individuals are more intelligent than poor individuals (Chapter 12). Population crashes may therefore have contributed to the evolution of our special intelligence by preferential survival of those who were more intelligent.

Mutation

As described in Chapter 3, the natural process of mutation produces roughly 1,000 new variants in each human child. Mutation provides the raw material on which the various forces described in this chapter act. Our special intelligence could not have evolved without mutation.

Chance

In all the hundreds of millions of years that complex organisms have existed on our planet, human-level intelligence evolved only once. Chimpanzees, gorillas, and orangutans didn't evolve human levels of intelligence, even though millions of years ago, humans and these species had common ancestors. Birds that mimic human speech haven't evolved special intelligence. Octopuses have had eight arms to efficiently manipulate objects for over 300 million years, yet have not built underwater cities. If the evolution of human-level intelligence were a normal progression in the evolution of all living things, then why didn't it evolve earlier, and why aren't there many species with human levels of intelligence?

I think at least part of the answer to these questions is that random chance played a major role in the evolution of our special level of intelligence. Key mutations may have occurred in just the right nucleotides, at just the right times, and in just the right individuals.

Almost nothing is known about such hypothesized chance events. Scientists do know, however, that about 66 million years ago, a large asteroid smashed into the Yucatán Peninsula in Mexico, causing the extinction of the dinosaurs and many other species. Although mammals coexisted with dinosaurs for millions of years prior to the asteroid, the absence of dinosaur competitors allowed mammals to diversify into many different and larger forms. Without the asteroid collision, there quite possibly would have been no apes and no humans.

Could chance events be solely responsible for the evolution of humans' special intelligence? I think this is unlikely. The evidence described in this chapter strongly indicates that other forces were involved. Without positive natural selection, it is improbable that the many new variants that slightly increased intelligence could have increased in frequency in our ancestors solely by chance. But if, as I suspect, the evolution of our intelligence was an especially lucky event, then our species is even more precious than we generally assume. I think it is therefore logical to expend exceptional effort to ensure that humanity, with our special skill, survives (126).

Summary

- Over about the last seven million years, since the time of the common human-chimpanzee ancestor, humans evolved a special level of intelligence that is unique among all species on the planet.
- Although the process that led to our special intelligence was undoubtedly complex and is still largely a mystery, three *facts* about the process are known:

  - Our special intelligence evolved through changes to our genomes.
  - Mutation created the new DNA variants that increased the intelligence of our ancestors.
  - On average, our more intelligent ancestors reproduced at higher rates than our less intelligent ancestors.
- Geneticists have identified at least several of the major forces that were likely involved in the evolution of our special intelligence (Table 13.2). There is considerable overlap among many of the forces.
- Differential reproduction is defined as differences in reproductive rates among people who carry different DNA variants.
- Natural selection acts by decreasing the frequencies of deleterious variants and increasing the frequencies of advantageous variants.
- Genealogical demography is the study of the numbers and characteristics of people's ancestors and descendants. Many people who lived in the past made no or only small contributions to the current gene pool, and others made large contributions.
- Polygyny, which means one male reproducing together with multiple females, was likely common throughout human evolutionary history.
- Changes in population size can affect the gene pool by altering the level of genetic diversity and the rates at which variant frequencies change.
- Migration leads to new combinations of variants.
- Infanticide, the killing of infants, was in the past common in many, perhaps nearly all, human societies.
- Assortive mating means that women and men who are similar in various traits, such as height or education level, tend to reproduce together.
- Population crashes, defined as relatively rapid drops in population size, were common in our past.
- Random chance events also played a role, perhaps a major role, in evolution of our special intelligence.

Chapter 14

THE PROBLEM: RECENT CHANGES IN EVOLUTIONARY FORCES

Starting about 12,000 years ago, life for our ancestors changed dramatically. Agriculture was developed; cities were built; written languages appeared. The Industrial Revolution, beginning in about the year 1800, accelerated the rate of change. Human populations exploded in size; lifespans doubled; fabulous new technologies were developed.

These changes improved people's lives enormously, but they also altered the forces that guided human evolution for millions of years. This chapter focuses on how the evolutionary forces introduced in the previous chapter have changed in relatively recent times, and how these changes are likely to affect the gene pool. As in the last chapter, I discuss the forces one by one, although in this chapter I invert the order from the previous one. I also add a new force, reproductive planning, which has arisen in just the last few decades. I continue to focus on intelligence, while not completely ignoring other traits.

It's important to keep in mind that the last 12,000 years are only a very small fraction (about 0.2%) of the seven million years since the human-chimp common ancestor. Very little is known about how evolutionary forces changed over nearly all of this time. As a baseline for discussing recent changes in these forces, I'm using what is known historically about early human agricultural societies and also what has been learned from studying hunter-gatherer populations.

Reproductive Planning

As described in Chapter 11, reproductive planning is expanding worldwide. Still, as of the time of writing, with the exception of NIPS, reproductive planning is utilized in only a small fraction of births, even in the wealthiest countries. In many developing countries, there is currently almost no reproductive planning at all. Therefore, to date, I gauge that the impact of reproductive planning on the gene pool has been minimal. If reproductive planning becomes more widespread in future, then this force may have an appreciable effect on the gene pool (Chapter 11), but not yet.

Chance

Random chance events probably occur today much as in the past. New variants and combinations of variants are certainly arising. Because the current human population is much larger than in the past, there is more opportunity for advantageous variants to arise, but less opportunity for population frequencies of these variants to change rapidly. With today's large populations, it would take many generations for even a decisively advantageous new variant to reach significant frequencies.

Even if chance events played a major role in the *evolution* of our special intelligence, I think it is less likely that chance will play a major role in the *maintenance* of our intelligence, at least over the next few generations. Other forces described in this chapter will probably have a much greater impact.

Mutation

The rate of germline mutation has probably not changed a great deal over the last seven million years. Germline mutation rates and mutation patterns in chimpanzees and gorillas are close to those in humans (1–4).

Especially after World War II, there was concern that radiation from nuclear weapons or from nuclear power plants would substantially increase human germline mutation rates. However, study of the incidence of single-gene disorders in the children of survivors of the atomic bombs dropped on Hiroshima and Nagasaki failed to detect increases in frequency (5–6).

Compared to currently available methods, the approaches used to detect changes in mutation rates in Japanese bomb survivors were crude. Today, genomes of parents and children can be directly sequenced to obtain much more sensitive measures of germline mutation rates. Although I know of no evidence for an increase in baseline mutation rates in the general population in recent decades, it is conceivable that small increases have occurred, due perhaps to all the chemicals in our modern world that were absent in the past. Since many more mutations are deleterious than advantageous (Chapter 3), any increase in the germline mutation rate would be detrimental to the gene pool.

One influence on mutation, however, that we know has changed substantially in recent decades is age at reproduction. In at least many nations, the age at which people reproduce is significantly higher today than in the past, and this is especially true for those who continue their education into college (7–8). As mentioned in Chapter 3, germline mutation rates (at least for nucleotide substitutions and short deletions and insertions) increase steadily with parental age (9–10). As an example, the number of nucleotide-substitution mutations in the average sperm cell increases by about one-third between 20-year-old and 30-year-old fathers. This increase in germline mutation due to increasing age at reproduction is adversely affecting the gene pool.

Population Crashes

Today, population crashes are considerably less frequent than in the past. Famines, wars, and pandemics certainly still occur, but they are much less frequent and involve much smaller fractions of populations (11–12). Although the 2020–2022 COVID-19 pandemic undoubtedly qualified as a major disaster, causing severe economic disruption and killing about 20 million people, the fraction of the world population that died was only about 0.25%. This small fraction is dwarfed by the approximately 90% of the Native American population who died from infectious disease introduced by Europeans in the 16th–18th centuries, or the 50% of Europeans who died during the 14th-century Black Death (Table 13.5).

If, as I surmised in the last chapter, population crashes tended to favor survival of the healthier and more talented, then the reduction in crashes, as beneficial as it is to humanity, is detrimental to the gene pool. Those who are less healthy and talented are more likely today to survive and reproduce than in the past.

Assortive Mating

Assortive mating for intelligence has been strong in at least some countries for centuries (13–14). However, the ongoing trends throughout the world for greater numbers of women to attend college and to enter the workforce may be increasing assortive mating for intelligence by providing more opportunity for men and women of higher intelligence to interact and reproduce together (14–17). An increase in assortive mating for intelligence would broaden the distribution of intelligence in a population and increase the numbers of people with both high and low intelligence, but would not change average population intelligence (Figure 13.3). I conclude that assortive mating for intelligence is likely increasing, but don't think that this change by itself is having a big effect on the number of deleterious variants in the gene pool.

Infanticide

Infanticide is one of the evolutionary forces that has clearly changed enormously in recent times. Throughout the world today, infanticide is greatly reduced compared to the past. In developed nations, it has been nearly completely eliminated.

The near elimination of infanticide is a wonderful advance for humanity, but this change has not been positive for the gene pool. Many more babies with congenital abnormalities and with poor health survive today than in the past. Many of these survivors reproduce, thereby passing on their deleterious variants to future generations. These are not pleasant thoughts, but they are genetic reality.

Migration

Migration has always been a part of human behavior (see previous chapter), but modern methods of transportation have dramatically increased the rates of migration and the distances traveled. As just one example, prior to the Africa-America slave trade, there were almost no individuals with (recent) African ancestry in America. Today, at least 15% of people in the US (18–19) and at least 50% in Brazil (20–21) have relatively recent African ancestry.

Family trees also provide evidence that migration has increased over time. In a study involving the construction of family trees for millions of individuals, mostly from the US and northwestern Europe, it was reported that the distance between birthplaces of married couples increased approximately 10-fold from 1800 to 1950 (22). Relatedness between spouses also decreased as the distance between birthplaces increased.

The vast ongoing movement of people from rural areas to cities in recent decades is further evidence for increased migration. In 1900, about 87% of people throughout the world lived outside of cities whereas in 2024, the number was 42% (23–24). When people lived their entire lives near or within small rural towns and villages, they had little choice but to reproduce with people from the same area to whom they were sometimes distantly related. But when people move to large cities, the chance that they will mate with a related person is diminished (25–28).

In plant breeding, crosses between two inbred strains usually result in offspring that have higher agricultural yields than either parental strain (29). This phenomenon is called hybrid vigor, or heterosis. Although it does not appear to be a major factor, there is evidence that heterosis also occurs in humans (25, 30–34). On average, the health, height, and talents of children increase as the ancestral distance between the parents increases. It has even been suggested that heterosis is responsible for the Flynn intelligence effect (Chapter 12) (35).

I judge that the increase in migration is one of the few recent changes in evolutionary forces that has been positive for the gene pool. Increased rates of migration have reduced inbreeding and the rates of recessive disease, and increased the combinations of variants upon which evolutionary forces may act.

Population Size

A plot of world human population size over time is shown in Figure 14.1. Prior to the advent of agriculture, the human population may have been only a few million or even fewer (36). Agriculture allowed the world population to grow, hitting 10 million about 8,000 years ago, and 100 million about 2,500 years ago (37–39). This early growth was sluggish, however, compared to the population explosion that took place over about the last 250 years. World population has gone from about 0.8 billion in 1775 to 8 billion today. So, the human population grew 10-fold from 10 million to 100 million over 5,500 years, and 10-fold from 0.8 billion to 8 billion in just the last 250 years.

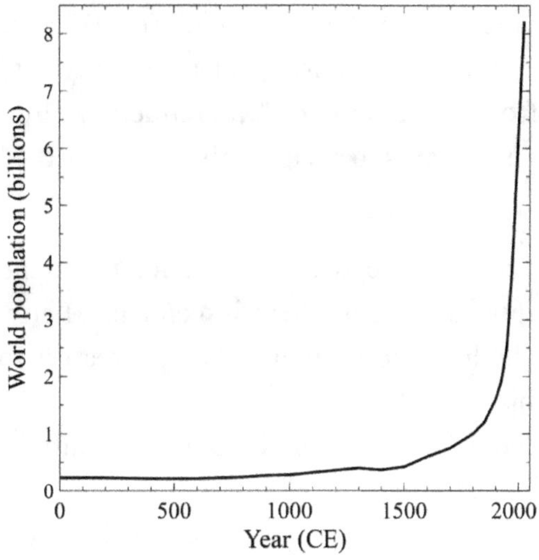

Figure 14.1 Human population growth
(adapted from references 37–39)

Like increased migration, increased population size has contributed to reductions in inbreeding and incidences of recessive disease (25–27). Population growth has also increased the amount of raw material on which natural selection and other evolutionary forces can act. Both the number of rare variants and the number of combinations of variants are increased.

Polygyny

Although polygyny still exists today, it is greatly reduced compared to the past (40–41). Polygamy is illegal throughout the world except in most Muslim-majority countries and in some African countries. Whereas nearly all hunter-gatherer societies practiced polygyny, very few men these days, even in the Muslim world, have more than one wife. Certainly, some men divorce and remarry. Some men also father children outside of their marriages. But the blatant polygamy/harem practices of the past have almost entirely disappeared.

I think the great majority of people, myself included, consider the reduction in polygyny a positive human development. But I also think the reduction in polygyny has had a negative impact on the gene pool. In the past, when polygyny was common, there was robust competition among men for wives. It was the dominant, leading men who had the most wives and produced the most children. When men with lower social status had few or no children, deleterious variants were reduced or eliminated from the gene pool. The reduction in polygyny has allowed deleterious variants to accumulate.

Genealogical Demography

In the previous chapter, I focused on the ancestors of people living today; in this chapter, I focus on their descendants. I attempt to answer the question: who living today will make the greatest contributions to future gene pools?

Obviously, people who are childless make no contributions to future human gene pools. People are childless for many reasons, including infertility, homosexuality, and a conscious decision not to have children. Because of substantial improvements in methods for and access to birth control and abortion, people today can much more easily avoid having children than in the past. Rates of childlessness have changed substantially over time and also vary considerably today from country to country (42–43). However, rates of 20%–30% for women are not unusual, and rates for men are higher—in some countries more than 1.5 times the female rate (44). In at least many parts of the world, rates of childlessness are currently increasing (42, 45). In a 2023 survey from the US, 47% of adults aged 18–49 said they are unlikely to ever have children, up from 37% in 2018 (46).

For those who do have children, their expected numbers of descendants can be gauged by modeling. In Tables 14.1 and 14.2 are shown the expected average numbers of descendants in various future generations as a function of the number of children in the current families. To generate these numbers, I set childlessness at 30%. For Table 14.1, I assumed that each descendant who reproduces has an average of 2.8 children, keeping the total population size constant. For Table 14.2, I assumed that each descendant who reproduces has the same number of children as the people in the current generation. For instance, if a person has three children, then each of those children who reproduces will also have three children, and all their descendants who reproduce will also have three children.

Table 14.1

Projections of average numbers of descendants of people living today as a function of family size (each reproducing descendant has the same number of children)

Generation	Number of Children				
(Years into Future)	1	2	3	5	10
1 (30)	2	4	6	10	20
2 (60)	4	8	12	19	38
3 (90)	8	15	23	38	75
10 (300)	837	1,673	2,510	4,183	8,367
20 (600)	0.70M	1.40M	2.10M	3.50M	7.00M

The first generation are the grandchildren of parents living today; the second generation are the great-grandchildren, etc. Numbers are rounded to the nearest whole number except when they approach or exceed one million (M).

My models show, not surprisingly, that people living today who have the most children are those who will on average contribute the most to future gene pools. For instance, even when family size is not transmitted across generations (Table 14.1), a person today who has 10 children

Table 14.2
Projections of average numbers of descendants of people living today as a function of family size (each reproducing descendant has the same number of children as the original individuals)

Generation	Number of Children of People Living Today				
(Years into Future)	1	2	3	5	10
1 (30)	1	3	6	18	70
2 (60)	0	4	13	61	490
3 (90)	0	5	28	214	3,430
10 (300)	0	58	5,004	1.38M	2.82B
20 (600)	0	1,673	8.35M	0.38T	0.80MT*

The first generation are the grandchildren of parents living today, the second generation are the great-grandchildren, etc. Numbers less than one million are rounded to the nearest whole number. M is million; B is billion; T is trillion. *0.80 million trillion

is still expected to have 10 times as many descendants as a person with only 1 child. And when family size is transmitted across generations (Table 14.2), the differences are greatly magnified. A person today who has 10 children is expected to have 686 times the number of descendants as a person with 2 children after just 3 generations. After 10 generations, the person with 10 children is expected to have 49 million times the number of descendants as the person with 2 children. Results from my models match fairly well what is known from real human families (47–49). So, the answer to the question I posed at the beginning of this section—namely, who living today will make the greatest contributions to future gene pools—is the people with the most children, and especially those with the most children whose descendants also have large families.

Although family size is not perfectly transmitted from one generation to the next, there is evidence that it is transmitted to a modest degree (50–55). There are also groups today who, for cultural or religious reasons, favor large families. If (and this is a very big IF), these groups maintain their preference for large families over many generations, they will come to dominate future gene pools (49).

I also conclude that the probability of a person's line of descent terminating decreases as their number of children increases. The study of four generations of people living in northern Sweden from 1885 to 2007 found that people in the first generation who had 1, 2, or 3 children, had a 55%, 22%, and 13% chance, respectively, of not leaving any great-grandchildren (56).

Another important factor in determining the number of descendants is the age of the person at the time of reproduction. The numbers in Table 14.3 show how the numbers of descendants of early reproducers (reproduction at age 20) differ from those of late reproducers (reproduction at age 33). After just a single century, an early reproducer has four times the number of descendants as a late reproducer. There is evidence from real human families that age at first birth is transmitted to a modest degree from one generation to the next (49–51, 57). And the trend toward intergenerational transmission of age at first birth may have accelerated in recent decades, especially for women. In the past, women rarely attended college, but today in developed nations, women frequently attend college and delay childbirth until their education is complete. In contrast, women who do not attend college often begin reproducing shortly after completing secondary school. Since educational attainment is significantly heritable, intergenerational transmission of age at first birth may be increasing. Under the simple model of Table 14.3, late-reproducing individuals would need to have 3.2 children for every 2 children in the early-reproducing individuals to make the total number of descendants equivalent.

Demographers have identified a transition, which nearly all nations outside of Africa are currently undergoing or have already completed, from relatively large families and rapid population growth, as shown in Figure 14.1, to relatively small families, increased age at reproduction, and population plateau or even decline (58–59). If current trends continue, the populations of Brazil, Japan, Italy, and China, for example, are projected to decline by 23%, 37%, 38%, and 55%, respectively, by the year 2100 (60).

Table 14.3
Numbers of descendants of earlier versus later reproducers

Years	Reproduction at Age 20	Reproduction at Age 33	Ratio of the Two Groups
100	32	8	4
200	1,024	64	16
300	32,768	512	64

Numbers in the table assume that all people have exactly two children, and that early or late reproduction continues in the two groups generation after generation.

This demographic transition affects the gene pool in at least three ways. First, as mentioned above, because germline mutations increase with age, increased age of parents at the time of reproduction will increase the numbers of mutations in the children. Second, the smaller families will reduce the effectiveness of natural selection (61). And third, those relatively rare families today with especially large numbers of children will make relatively greater contributions to the future gene pool than in past decades when large families were common. I rate these changes as negative for the gene pool.

Natural Selection

Over about the last 200 years, average human lifespan has approximately doubled (58, 62–63). While reductions in both maternal mortality and overall adult mortality have contributed to this doubling, by far the largest factor has been massive drops in infant and child mortality (58, 64–67). In nearly all preindustrial human societies, both hunter-gatherer and agricultural, 30%–50% of all newborns died before reaching reproductive age. Today, in wealthy nations, the rate is well under 1%. Little is known about mortality rates prior to the advent of anatomically modern humans, but pre-reproductive mortality rates in wild chimpanzees and gorillas may be even higher than in preindustrial human societies (64, 67–69).

It's therefore likely that for nearly all of the seven million years between the human-chimp common ancestor and the present, a large fraction of newborns did not survive to reproductive age. Famine, disease, infanticide, violence, storms, accidents, and predators killed many children. When such a large fraction of offspring died, the odds against those who were unhealthy or disabled were high. For the most part, it was those who were above average in health, vitality, and intelligence who survived and reproduced. Although it is indeed a marvelous advance for humanity, the huge drop in pre-reproductive mortality over the last two centuries has *profoundly* altered natural selection. It is difficult to overstate the magnitude and significance of this change. Enormous numbers of people who would not have survived in the past are now surviving and reproducing. There are many examples of this phenomenon, but I will present only four: people with type 1 diabetes, severe bacterial infections, congenital heart disease, and deafness.

The onset of type 1, also known as juvenile, diabetes is usually between 4 and 14 years of age. In this disease, the cells in the pancreas that produce the hormone insulin are destroyed by autoimmunity. Without insulin, blood sugar rises unchecked, and in the past invariably resulted in death (70–71). However, after insulin started to be produced pharmaceutically in 1922, children with type 1 diabetes began to survive. Today, nearly all affected children in developed nations survive to reproductive age. Reproduction in type 1 diabetics is reduced compared to unaffected individuals, but is still substantial (72–74).

In the past, huge numbers of people died from bacterial infections. As just one example, bubonic plague caused by the bacterium *Yersinia pestis* is estimated to have killed 100 million people in the years 541–543 and 200 million in 1347–1351 (75). Today, due to improved sanitation, as well as vaccines and antibiotics, the annual worldwide number of deaths from plague, despite massive population growth, is only about 2,000. Our immune systems, which protect us against bacterial and other infections, are complex and like all other biological systems are susceptible to degradation by mutation (76). Although in the past bacterial infections undoubtedly took the lives of many healthy and talented people, those with weakened immune systems were particularly susceptible. Today, most individuals with weakened immune systems survive bacterial infections, and many reproduce.

Congenital heart disease affects about 1 in 100 newborns (77–78). Many types of heart defects occur, including holes in the chambers of the heart, faulty valves, and abnormal plumbing. In the past, nearly all such babies succumbed early in life. Modern surgical techniques, however, can repair many of these defects and allow the babies to survive. I'm amazed at what can be done these days to help these children. A significant fraction of congenital heart disease is explained by rare DNA variants (79–82), and complex inheritance likely explains many other cases (83). Although fertility likely depends upon the severity of disease, reproduction overall in congenital heart disease patients appears to be only modestly reduced compared to unaffected individuals (84–85).

Substantial hearing impairment (deafness) affects about 1 in 1,000 newborns (86–87). Most cases of newborn deafness are single-gene disorders, with over 200 different genes involved. Deafness is inherited in recessive, dominant, and X-linked patterns; recessive inheritance predominates. Although data are sparse, it appears that in the past, reproduction by deaf individuals was low (88–89). The introduction of sign languages and dedicated schools for the deaf over the last few centuries markedly improved the lives of deaf individuals. Recent rates of reproduction of deaf individuals have been reported to be about 60% those of hearing individuals in developing nations (90–91), and about 90% in developed nations (89, 92). In addition, probably in large part because of schools for the deaf, deaf individuals practice a high level of assortive mating. There is evidence that this assortive mating has increased the incidence of deafness (93–95).

Readers need to understand that I am not attempting, here or elsewhere in the book, to denigrate those with poor health or disabilities. No one chooses to have diabetes, bacterial infections, congenital heart disease, or deafness. Also, most people with these conditions can have healthy children through careful reproductive planning. Nevertheless, the unpleasant reality is that deleterious variants that in the past would have been removed by natural selection are now accumulating in the gene pool (96–97). This is true not only for the four examples described above, but for *many* other health problems and disabilities.

Although natural selection has been substantially relaxed in our modern societies, it has not been abolished entirely. Many aspects of natural selection are still active today. For example, *all* embryos and fetuses that have lost one of the two autosomal chromosomes spontaneously abort. All embryos and fetuses that entirely lack essential proteins due to the presence of loss-of-function variants in both gene copies spontaneously abort. People with especially severe health problems or disabilities often do not reproduce (98–102). Any new variant that causes death before reproductive age or results in sterility is immediately removed from the gene pool in that generation.

Differential Reproduction

In the previous chapter, I pointed out that in order for our special intelligence to have evolved, the more intelligent of our ancestors must have reproduced at higher rates than the less intelligent. Furthermore, because mutation continually erodes our collective intelligence, the more intelligent among us today must reproduce at higher rates than the less intelligent just to *maintain* humanity's current average level of intelligence. A critical question, then, is: Are the more intelligent currently reproducing at higher rates than the less intelligent?

Abundant evidence exists that in preindustrial societies, high-status men—as indicated by income, wealth, authority, occupation, and education—reproduced at higher rates than low-status men (103–110). Although status is certainly not the best measure of intelligence, higher-status men then, as now, are on average more intelligent than lower-status men (Chapter 12). In preindustrial societies, therefore, more intelligent men likely out-reproduced less intelligent men. But beginning with the Industrial Revolution, the trend for higher-status men to out-reproduce lower-status men, which had probably been in effect for millions of years, slowed or in some cases reversed (106, 108–110).

Almost nothing is known about the influence of talent on women's reproduction prior to the Industrial Revolution. It is certainly plausible that higher-intelligence women left more descendants due to greater success in raising children who survived to adulthood and in attaching themselves to high-status men, but this remains unproven.

In the 20th century, it became possible to relate reproduction to better measures of intelligence than status—namely, educational attainment and results of intelligence tests. The quality of population data also improved, allowing reproduction to be more accurately assessed. The two most important parameters for reproduction are total numbers of children produced over a lifetime and parental age at birth of the children.

Many studies, mostly nation-based, have now been published on the relationship between total lifetime reproduction and educational attainment or intelligence testing. These studies have consistently concluded that in most of the 20th century, in developed nations, women with higher intelligence gave birth to substantially fewer children than women with lower intelligence. This result holds for the US (105, 111–117), Canada (118–119), Europe (120–129), Japan (130–131), Taiwan (132), and South Korea (133).

For men, the data are sparser and the conclusions mixed. In some countries, including especially the US, more intelligent men had fewer children than less intelligent men, although the gap was not as large as for women (112, 117, 125, 127, 130, 133–135). In other countries, mid- and high-intelligence men had about the same total number of children, but low-intelligence men had substantially fewer children (125, 131, 136–139). Some representative results for both men and women in the US are shown in Figure 14.2, and for women in Japan in Figure 14.3.

Another important finding, which is universal among all countries studied and applies to both women and men, is that on average, over the last several decades, lower-intelligence individuals began reproducing earlier in life than higher-intelligence individuals (see, for example, 116, 118, 129, 135, 137). This is not surprising since those who undertake higher education typically postpone having children until their education is complete. As shown in Table 14.3, over multiple generations, those who reproduce earlier in life will make greater contributions to future gene pools than those who reproduce later.

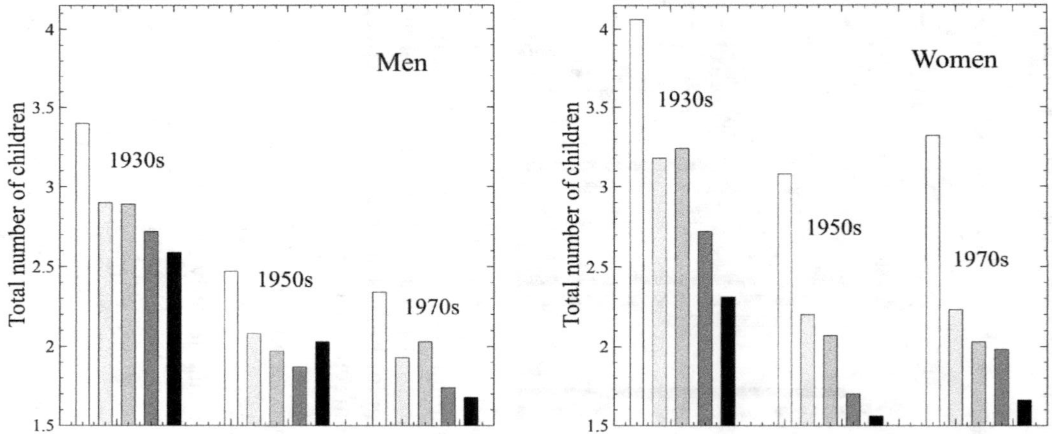

Figure 14.2 Total number of children versus educational attainment in the US
Results for men are on the left and for women on the right. Shown are three birth cohorts: people born in the 1930s, 1950s, and 1970s. Bar shading, from lightest to darkest, indicates level of educational attainment: less than high school, high school, some college, bachelor's degree, and graduate degree (adapted from reference 117).

There is also evidence that lower-intelligence individuals more often produce large families (three or more children) than higher-intelligence individuals (119, 137, 140–142). As noted in the section above on genealogical demography, people with three or more children are likely to make greater contributions to the future gene pool than people with one or two children.

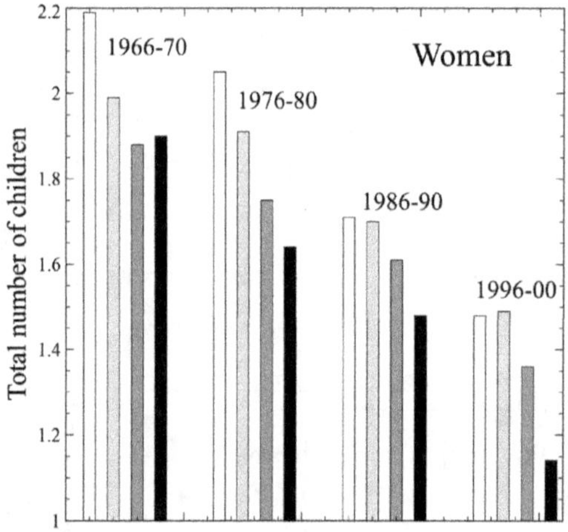

Figure 14.3 Total number of children for women versus educational attainment in Japan for the indicated dates
Bar shading, from lightest to darkest, indicates level of educational attainment: junior high, senior high, junior college, university (adapted from reference 130, table 2).

There are at least two significant limitations to the conclusions of the previous few paragraphs. One is that, for the most part, usable demographic data are available only for developed nations. The situation in developing nations is unknown. Even for China, few results are available (143–144). A second limitation is that data on completed reproduction is only available for women who have reached the age of about 45 (born in 1980 or earlier), and men about 65 (born in 1960 or earlier). Results for people who are reproducing today are unavailable. Demographers have taken current reproduction rates and extrapolated to total lifetime fertility (see, for example, reference 115), but I don't know the accuracy of this extrapolation.

Results for differential reproduction as a function of intelligence are not all gloomy; there are some encouraging signs. As noted above, the gap in total number of children between higher- and lower-intelligence men is considerably smaller than between higher- and lower-intelligence women. Also, over the last few decades, the reproductive gap between higher- and lower-intelligence women has narrowed appreciably in some, but not all, developed nations (125, 127–128, 131, 133, 141, 145).

Nevertheless, despite these encouraging signs, available evidence for differential reproduction indicates that at least in developed nations, for the trait of intelligence, the gene pool took a hit in the 20th century. More intelligent individuals had fewer children, reproduced later in life and were less likely to have large families than less intelligent individuals. DNA variants that are deleterious for the trait of intelligence accumulated in gene pools. This decline appears to be continuing in the 21st century in at least many parts of the world.

The Net Effect of Changes in Evolutionary Forces: The Gene Pool Problem

My estimation of the effect of relatively recent changes to each of the evolutionary forces discussed in this chapter on the trait of average innate intelligence is summarized in Table 14.4. Note that these results are for *average* intelligence of populations; they do not apply to the distribution of intelligence (which has probably broadened in recent years through increases in assortive mating). Of the 12 evolutionary forces, I rate two—migration and population size—as modestly beneficial to the gene pool; three— assortive mating, chance, and reproductive planning—as neutral; and the remaining seven—genealogical demography, infanticide, mutation, polygyny, population crashes, natural selection, and differential fertility—as detrimental to the gene pool. Clearly, by this analysis, negative changes outnumber positive changes.

Furthermore, it is very unlikely that all 12 forces have equal impact. I subjectively rated impact in the second column of Table 14.4. The changes to the forces that I think are having the greatest impact—mutation, polygyny, population crashes, natural selection, and differential fertility—are all negative. I conclude, therefore, that deleterious variants for intelligence and many other traits are currently accumulating in the gene pool. This is the gene pool problem.

Table 14.4
Effect of recent changes in evolutionary forces on average innate intelligence

Force	Impact on Gene Pool	Effect on Gene Pool: Positive	Neutral	Negative
Migration	+	X		
Population size	+	X		
Assortive mating	+		X	
Chance	+		X	
Reproductive planning	+		X	
Genealogical demography	+			X
Infanticide	+			X
Mutation	++			X
Polygyny	++			X
Population crashes	++			X
Natural selection	+++			X
Differential fertility	+++			X

This conclusion is so important that I will restate it using different words. The relentless process of mutation continually generates new deleterious variants that erode human health, intelligence, and other abilities. In the past, natural selection and other evolutionary forces efficiently kept these deleterious variants in check. These forces were also effective in increasing the frequency of rare advantageous variants. But beginning about the year 1800, impressive human advances such as reductions in infant and child mortality have had the unintended and largely unrecognized effect of altering the evolutionary forces such that the gene pool is degrading and our average health and abilities are dropping. Our special intelligence is fragile (146–147). It took millions of years to evolve, but can be significantly damaged in just a few generations.

The deleterious variants with large effects are not the only concern, or even the greatest concern. Many of the dominant deleterious variants that cause severe disease or disability are removed from the gene pool even today within one or a few generations. I think the greatest concern are the deleterious variants that clearly diminish our health and abilities, yet have small enough effects that they escape removal from the gene pool via relaxed natural selection and other changes to the evolutionary forces.

There is another approach, other than analyzing changes in evolutionary forces, that has been used to attempt to determine whether average, innate intelligence is changing over time. This approach involves the use of polygenic indices to estimate the intelligence of people born at different times. Very likely, the best study to date using this approach was carried out using nearly the entire population of Iceland (148). The authors of this study concluded that for Icelanders born from 1915 to 1985, indices for educational attainment dropped consistently and significantly with time. A second study from a much smaller US cohort also concluded that polygenic indices for educational attainment were dropping over a period of decades (149).

Many of the concepts described in this chapter, although updated by me, and although based on much more evidence than was previously available, are not new. Many other geneticists, perhaps most notably Francis Galton (150), Charles Darwin (151–152), Ronald Fisher (153), Hermann Muller (154), James Crow (155–156), William Blau (157), and Michael Lynch (158–159), have previously described the gene pool problem—in the case of Galton and Darwin, over 100 years ago. None of these men were/are perfect, but they all accomplished a great deal in their lives. Anyone who thinks these men were monsters should read an article by Muller on the gene pool problem (160), and also read obituaries for Crow (161–162). Why do geneticists keep bringing up the gene pool problem decade after decade? I think it is because this problem is real and needs to be addressed.

What Is the Impact of Deleterious Variants Accumulating in the Gene Pool?

If, as I and other geneticists have concluded, deleterious variants are currently accumulating in the gene pool, then *critical* questions are what is the rate of degradation and how long before we begin to observe noticeable effects? Unfortunately, I don't think anyone can yet confidently answer these questions. But at least we can begin to consider them.

A number of researchers have attempted to measure the rate of decline of average, innate intelligence by analyzing differential fertility as a function of intelligence (112, 132, 163–164) or by using polygenic indices (149). Estimates generally are in the range of 0.5%–1% decline per generation. I stress that these are initial, crude estimates, and that they should not be taken as scientific fact. Still, even a 1% decline in average innate intelligence could have an appreciable effect, and if these trends continue generation after generation, the combined effect would be ruinous.

I don't think anyone knows what the first signs of a gradual decrease in average intelligence would be, but I suggest it might be effects such as declines in the effectiveness and efficiency of schools, businesses, infrastructure, and governments. Many tasks would take longer to accomplish. More mistakes would be made in nearly all activities. Smaller problems would become larger problems, and larger problems would become insurmountable. I am particularly fearful for democracy. People with lower intelligence are more easily confounded by demagogues, and it takes only 51% of votes to win elections.

A case could be made that we are already witnessing the effects of a reduction in average intelligence. I think this is a distinct possibility, but I am far from certain. We need to keep in mind that humans have *always* faced daunting problems. Just because humanity faces big challenges today does not mean these problems are due to a drop in average intelligence.

If deleterious variants are accumulating in the gene pool, then why do impressive scientific and technological advances continue to be made? Part of the answer, I think, is that the momentum built up especially since the start of the Industrial Revolution has not yet dissipated. Another factor may be that due to assortive mating, our societies continue to have many highly intelligent people, even as average intelligence is dropping. I think it is possible that our rapid scientific and technological progress has at least partially masked the effect of a drop in average intelligence.

The gene pool is probably degrading at different rates in different parts of the world. Perhaps it is degrading the fastest in the most developmentally advanced nations. However, even the poorest regions of the world have seen large changes in the evolutionary forces. Infanticide, polygyny, and infant and child mortality have dropped everywhere. Natural selection has been substantially relaxed. I'm convinced that the gene pool problem is global.

I think it would be extremely foolish to ignore the gene pool problem. Complacency could be disastrous. In Chapter 12, I expressed the opinion that humanity has just barely evolved sufficient intelligence to form advanced civilizations. If this is correct, then even small drops in intelligence could be devastating. And our societies certainly cannot tolerate continued drops in average intelligence generation after generation. If my analysis is correct, and if we want to maintain our advanced human societies, then we *must* address the gene pool problem. It's also clear that the longer we wait to tackle this problem, the harder it will be to solve (158–159). We need to face the gene pool problem and begin to devise solutions *now*.

Summary

- In the past, the various evolutionary forces described in Chapter 13 kept mutation at bay by efficiently removing new deleterious variants from the gene pool.
- As a result of relatively recent changes to the evolutionary forces (summarized in Table 14.4), deleterious variants are now accumulating in the gene pool. Average health and abilities of our species are declining. This is the gene pool problem.
- Many other geneticists have recognized the gene pool problem in modern times, beginning with Francis Galton and Charles Darwin over 100 years ago.

- The rate at which the gene pool is degrading is unknown, but it would be dangerous to ignore this problem.
- If my analysis is correct, and if we want to maintain our advanced human societies, then we *must* solve the gene pool problem. We have no choice.

Chapter 15

HISTORY OF GENE POOL MANAGEMENT

In the previous chapter, I defined the gene pool problem. In Chapters 16 and 17, I cover possible solutions to this problem. In this chapter, I present a brief history of gene pool management as an introduction to Chapters 16 and 17.

I define gene pool management as active steps taken by human societies to prevent single-gene disorders and to control the number of deleterious variants in the gene pool. Some gene pool management steps have likely been employed in the human ancestral line for millions of years.

I'm a geneticist, not a historian. The vignettes presented in this chapter provide only a taste of the history of gene pool management. My strong guess is that I have missed many examples of gene pool management for which historic evidence exists, and that an even larger number of examples from the past have gone unrecorded and will never be known to us.

Except for the discussion of flaws in the Eugenics Movement, I do not offer personal opinions in this chapter. My presentation of this history does not in any way indicate my approval of the ideas or events.

Consanguineous Matings

Consanguineous matings are reproduction between closely related individuals. Because nearly all individuals are heterozygous for a few recessive deleterious variants with high penetrance (Chapter 5), and because closely related individuals often inherit the exact same deleterious variant from a common ancestor, consanguineous matings produce offspring with increased rates of recessive single-gene disorders (1–3). The closer the relationship between the parents, the more likely it is that the children will be homozygous for deleterious variants. The family tree of a child who is the offspring of a first-cousin marriage and who is homozygous for a recessive deleterious variant is shown in Figure 15.1.

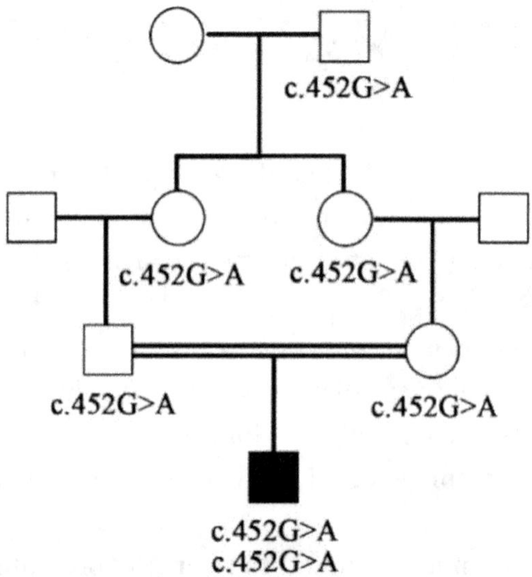

Figure 15.1 Recessive disease in a child of a first-cousin mating

A rare deleterious variant in the *DHCR7* gene (c.452G>A (p.Trp-151Stop)), causative for the recessive disease Smith-Lemli-Opitz syndrome, is inherited by both daughters of the great-grandfather, and also by the children of the daughters, who are first cousins. A son born to the first cousins inherits the deleterious variant from both his father and his mother and is consequently affected. The double line between the cousins indicates that this is a consanguineous mating.

Rates of consanguineous matings vary dramatically across the world (4–5). In many parts of the Muslim world, marriages between first cousins are common, comprising a large fraction of all marriages. In parts of India, especially in the past, uncle-niece marriages were preferred. In places with high rates of consanguineous marriages, the incidence of recessive single-gene disorders is much higher than in regions with low rates (6–9).

In Europe, the Catholic Church banned marriages between cousins beginning in about the year 600 (10). Consequently, first-cousin marriages are relatively rare in predominantly Christian countries. In the US, first-cousin marriages are prohibited by law in 24 of the 50 states (11).

Incest is a rare, extreme form of consanguineous mating. It is often defined as a parent-child or sibling mating. A large fraction of the offspring of incestuous matings have significant health problems and/or disabilities. In a series of studies involving 190 such children, 83 (44%) had severe health problems, and about one-quarter were intellectually disabled (12). More recent studies have confirmed the high rates of disease and disability among children of incestuous matings (13–15).

Incest has been taboo and/or formally prohibited in nearly all human societies for which records are available (12). It was also probably avoided in the human lineage for millions of years before that. Incestuous matings are rare even in chimpanzees. Chimp females, when they reach reproductive age, usually leave their home group and join another group (16–18). Control of consanguineous matings and avoidance of incest are simple forms of gene pool management.

Classical Greece

Gene pool management was widely considered by classical Greek philosophers (19). About 2,400 years ago, Plato, in Book V of *The Republic*, after briefly discussing the use of artificial selection in the breeding of birds, dogs, and horses, had this to say about *human* gene pool management (20):

> It follows from what has already been granted that the best of both sexes ought to be brought together as often as possible, and the worst as seldom as possible, and that the issue of the former unions ought to be reared, and that

> of the later abandoned, if the flock is to attain first rate excellence. . . .

and

> These officers (appointed to care for children) will take the children of good parents and place them in the general nursery under the charge of certain nurses, living apart in a particular quarter of the city: while the issue of inferior parents, and all imperfect children that are born to the others, will be concealed, as is fitting, in some mysterious and unknown hiding place.

Aristotle in *Politics* wrote (21):

> As regards whether to expose or nourish what is born, let the law be to nourish nothing that is defective.

Although gene pool management was an active topic of discussion in classical Greece, it is not clear to what extent it was actually practiced. There is evidence, however, that it was at least occasionally practiced. Abandonment (often called "exposure") of infants was mentioned by several Greek authors (22). In describing the militaristic Spartan society, the historian Plutarch wrote (23):

> The father of a newborn child was not entitled to make his own decision about whether to rear it, but brought it in his arms to a particular spot termed a lesche where the eldest men of his tribe sat. If after examination the baby proved well-built and sturdy they instructed the father to bring it up, and assigned it one of the 9,000 lots of land. But if it was puny and deformed, they dispatched it to what was called "the place of rejection", a precipitous spot by Mount Taygetus, considering it better both for itself and the state that the child should die if right from birth it was poorly endowed for health or strength.

Prussian Long Guys

King Friedrich Wilhelm I of Prussia, who reigned from 1713 to 1740, really liked tall soldiers. He scoured Europe for the tallest men he could find, sometimes resorting to kidnapping, and assembled them into a special regiment, numbering about 3,000, based in Potsdam, near Berlin. He also recruited tall women to marry the tall soldiers so that his regiment, often called the Potsdam Giants, would be perpetuated. In his book *The Descent of Man,* Darwin mentioned the Potsdam Giants as the only example he knew of attempted human artificial selection (24). A club was founded in Potsdam in 1990 called Die Lange Kerls (tall guys). For a fee, these tall men will dress in 18th century military uniforms and parade at festivals throughout the world (25).

"Good Breeding"

People in England, especially in the 17th–19th centuries, were often described as having good or bad breeding (26–27). "Good breeding" was used to describe a person, usually of the upper class and well educated, with good manners and pleasant speech, whereas "bad breeding" meant a loutish, uneducated person. The philosopher John Locke used the term, as did a number of novelists, including Jane Austen. In *Pride and Prejudice* (1813), Austen described one of her characters, a Mr. Bingley, as a man with "so much ease, with such perfect good breeding."

The Eugenics Movement

The Eugenics Movement started in England in the late 19th century, reached its peak in the 1920s and 1930s, and largely ended with World War II. The movement was started by the English polymath Francis Galton (1822–1911). Galton made major contributions to fingerprinting, weather maps, and especially statistics (28). Later in his career, he became interested in the inheritance of human traits (29).

In 1883, Galton coined the term "eugenics" to describe the genetic improvement of humanity and other species (30). He was worried that more talented people were reproducing at lower rates than less talented individuals. Galton wrote (31):

> If a twentieth part of the cost and pains were spent in measures for the improvement of the human race that is spent on the improvement of the breed of horses and cattle, what a galaxy of genius might we create.

Starting with Galton, the Eugenics Movement grew and spread (32–35). Eugenics societies sprang up in the US, Sweden, Germany, and many other countries. Whereas Galton seemed to be more interested in increasing the reproductive rate of the most talented, most of those who followed emphasized reducing the reproductive rate of the least talented, mainly through involuntary surgical sterilization.

At its height, the Eugenics Movement had *many* prominent advocates, including Winston Churchill, Theodore Roosevelt, Alexander Graham Bell, H. G. Wells, John D. Rockefeller, Luther Burbank, and George Bernard Shaw. Up until 1954, one of the most prominent human genetics journals was titled *Annals of Eugenics*. Physicians were nearly always the ones who selected individuals for sterilization, and, of course, were the ones who performed the sterilizations.

Ultimately, about 30 of the (then) 48 US states passed mandatory sterilization laws, beginning with Indiana in 1907. In the US, the Eugenics Movement was aided by concerns about immigration and the costs to taxpayers of institutions that in those days housed many of the intellectually and psychiatrically disabled. Not everyone, however, supported the movement. The Catholic Church, in particular, was steadfastly opposed. A number of legal challenges were mounted against the American mandatory sterilization laws. In 1927, the Virginia sterilization law was adjudicated by the highest American court, the Supreme Court, in the case of *Buck v. Bell*. The Supreme Court upheld the Virginia law by a vote of 8–1. In the majority court opinion, Chief Justice Oliver Wendell Holmes wrote:

> It is better for the world, if instead of waiting to execute degenerate offspring for crime, or to let them starve for their

> imbecility, society can prevent those who are manifestly unfit from continuing their kind.

and

> Three generations of imbeciles is enough.

Total numbers of persons who were involuntarily sterilized will probably never be known with certainty, but some rough estimates by country are shown in Table 15.1. Germany is a special case that is discussed below. Altogether in the US, about 60,000 individuals were involuntarily sterilized, the great majority in the 1920s and 1930s (33, 35). Perhaps another 60,000 were sterilized in Sweden (36–38), and smaller numbers in other countries (38–43). Although not listed in Table 15.1, involuntary sterilizations were also carried out in Estonia, Finland, Iceland, Poland, Switzerland, and probably other countries (38, 44–47). Interestingly, mandatory sterilization was never practiced in England, the birthplace of the Eugenics Movement.

Table 15.1
Estimates of the numbers of individuals involuntarily sterilized during the Eugenics Movement by country

Country	Numbers
Germany	400,000
United States	60,000
Sweden	60,000
Japan	17,000
Denmark	10,000
Canada	4,000
Norway	4,000

Even excluding the Nazi atrocities, nearly everyone, including myself, has concluded that the Eugenics Movement was fatally flawed. The primary flaws as I see them are listed in Table 15.2.

Many of the leaders of the Eugenics Movement were racists, at least by today's standards. As was common in that period, they took for granted that leading individuals with ancestry from northwest Europe were superior to all other people on the planet. Advocates in the US considered high rates of reproduction among those at the bottom of society, as well as immigration, to be threats to the "American race."

Table 15.2
Fatal flaws of the Eugenics Movement

- Racism was a primary motivation.
- Understanding of human genetics was poor.
- Advocates used derogatory terms such as "moron," "lunatic," "parasite," "ballast," and many others to describe those they considered genetically flawed.
- The movement advocated and practiced mandatory sterilization.

Many in the Eugenics Movement had only a weak understanding of human genetics. They incorrectly assumed that behaviors like crime, prostitution, sloth, and alcoholism were single-gene traits with simple inheritance patterns (see, for example, 48). The best human geneticists of those days fairly quickly realized that such assumptions were incorrect (see, for example, 49), but these geneticists didn't immediately prevail.

Proponents of the Eugenics Movement used derogatory terms to describe those they considered unfit. These terms tended to demean people who were disabled and to decrease public sympathy for them. Compare the term "idiot" to the current favored term "intellectually disabled."

Finally, many (although not all) of the leaders of the Eugenics Movement advocated and helped to implement mandatory sterilization. The majority of those sterilized in the US were residents of institutions for the intellectually and psychiatrically disabled.

After World War II, the Eugenics Movement rapidly dissipated. There were relatively few involuntary sterilizations performed after the war. In contrast, *voluntary* surgical sterilizations, for purposes of birth control, increased substantially after the war. Some of these "voluntary" sterilizations were probably coerced to control population size, as in China, but do not appear to have been performed overtly to manage the gene pool (35).

Except perhaps in Germany, I don't think that the Eugenics Movement had any appreciable impact on even national gene pools, much less on the worldwide gene pool. The numbers of people sterilized were too small.

Nazi Germany

The Nazis in Germany in the 1930s and early 1940s carried ideas from the Eugenics Movement to a chilling and gruesome extreme. The Eugenics Movement meshed with Nazi and pre-Nazi German ideas of Aryan racial superiority. In 1933, the Nazis began with involuntary sterilization of individuals deemed to be genetically flawed. Between 1933 and 1939, an estimated 400,000 were sterilized (50). In 1939, the Nazis switched from sterilization to outright murder (50–53). About 90,000 are thought to have been killed with bullets or gas. Later during the war, gassing of those deemed genetically flawed was largely discontinued so that genocidal efforts could be redirected toward Jews and the Romani. However, another approximately 150,000 of the intellectually or psychiatrically disabled were killed through starvation or exposure to the elements.

Singapore

Lee Kuan Yew, served as Singapore's first prime minister (in office 1959–1990) and is considered by many to be the "founding father" of Singapore. Lee was an ardent advocate of gene pool management. In 1983, he said (54–55):

> Whilst we have brought down the birth rate, we have reduced it most unequally. The better educated the woman is, the less children she has.

and

> If we continue to reproduce ourselves in this lopsided way, we will be unable to maintain our present standards. Standards of competence will decline. Our economy will falter, administration will suffer, and society will decline.

In 1984, the Singapore government introduced a number of measures to attempt to increase the reproduction of more educated women and decrease the reproduction of less educated women (56–58). Tax breaks were given to more educated women with three or more children, and their children were given priority for entrance into the best schools. The government also tried matchmaking services and all-expenses-paid "Love Boat" cruises for educated people. Cash payments were offered to less educated women who were willing to be voluntarily sterilized.

These government measures turned out to be highly unpopular. Educated women resented attempts to encourage them to have more children. Very few less educated women were sterilized. The programs were so unpopular that they were abandoned by the government within a few years. They appear to have had no appreciable effect on the Singapore gene pool.

Nobel Prize Sperm Bank

It has been known for many years that donated sperm could be used to impregnate women through the process of artificial insemination (Chapter 11) (59–60). Because fertile men produce vast numbers of sperm, men can in principle father virtually unlimited numbers of children through artificial insemination and/or in vitro fertilization.

In 1980, an optometrist named Robert Graham started a sperm bank in the US using donors who were deemed by Graham to be especially talented (61). The openly declared purpose of the bank was to produce unusually intelligent children. Graham's Nobel Sperm Bank operated from 1980 to 1999, but was ultimately unsuccessful. Apparently, only three of the donors were actual Nobel Prize winners. Only about 200 children were born using sperm from the bank.

Although the Nobel Sperm Bank flopped, other sperm banks are active and thriving today (Chapter 11) (62–63). Sperm banks are an ongoing form of gene pool management.

Reproductive Planning

Enhanced knowledge and capabilities in human genetics have led to dramatic improvements in reproductive planning to avoid single-gene disorders, aneuploidies, and other health problems in children (Chapter 11). Reproductive planning is another active type of gene pool management.

- - -

I think the most important point of this chapter is that gene pool management is not a new human concept, but rather has been considered for thousands of years and has probably been practiced in simple forms for millions of years. It did not arise with the Eugenics Movement and the atrocities of the Nazis.

Another important point is that gene pool management is being actively practiced today. Incest and often other consanguineous matings are avoided. Various assisted reproductive technologies are being employed in a small, but significant fraction of births.

Summary

- Gene pool management is not a new concept. Incest avoidance has probably been practiced in the human lineage for millions of years.
- Classical Greek societies considered and implemented gene pool management.
- The Eugenics Movement started in England in the 1880s and for a few decades was popular in a number of countries, including the US. The movement was fatally flawed for a number of reasons (Table 15.2), including racism, poor understanding of genetics, and restriction of reproductive freedom.

- The Nazis in Germany in the 1930s and 1940s carried ideas from the Eugenics Movement to a horrific extreme. The Nazis sterilized about 400,000 people and murdered another approximately 240,000 people they deemed genetically flawed.
- Aggressive gene pool management efforts in Singapore in the 1980s failed.
- Some forms of gene pool management are being practiced today. These include avoidance of consanguineous matings, sperm donation, and reproductive planning.

Chapter 16

RACISM, ELITISM, AND OTHER RISKS

In this chapter, I cover some of the risks inherent in gene pool management programs, with an emphasis on racism and elitism.

Racism

It's important for readers to understand that the gene pool problem is not restricted to any one group or groups of people. Mutation occurs in everyone. Deleterious variants are accumulating in all societies. No group or groups of people can be correctly or justly blamed for the gene pool problem.

In the previous chapter, I listed racism as one of the fatal flaws of the Eugenics Movement (Table 15.2). I define racism as discrimination against a person solely because that person belongs to a particular group. The group may be based upon geoancestry, religion, language, or other factors. I have been adamantly opposed to racism my entire life, and I reject any proposed solution to the gene pool problem that is racist in nature.

Although I have written extensively in this book about the genetics of intelligence, I have not up to this point mentioned possible differences in average intelligence among different human groups, particularly among what are commonly called races. I have avoided this topic because it does not alter the major themes of this book, and because I think discussion of this topic is harmful (1–2). I don't deny that differences in average innate intelligence among human groups are genetically possible, but even if such differences exist, *many* individuals within groups with lower averages will have higher intelligence than many individuals within groups with higher averages (Figure 16.1). There are highly intelligent people within *all* human societies. All societies can benefit from gene pool management.

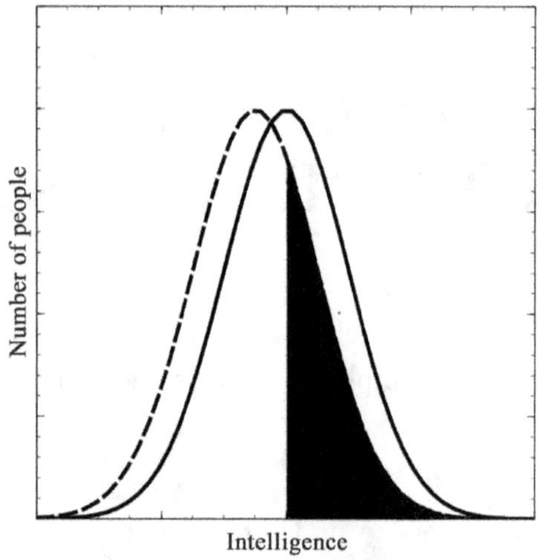

Figure 16.1 Normal distributions with slightly different averages

The dashed curve is a distribution with a slightly lower average than the solid curve. The shaded area represents the many individuals in the group with the lower average who have higher intelligence than the lower half of the group with the higher average.

Another reason why I think possible small differences in averages among groups are not a major factor is that migration is continually bringing together people with widely differing geoancestries. As history has shown countless times, when people live close to each other, they reproduce together (3–4). Especially in countries like the US, Brazil, Canada, and Australia, in which nearly all people are immigrants or the descendants of immigrants, large fractions of the populations already have relatively recent ancestors from two or more continents. In the US, about 20% of the population classifies themselves as Hispanic, and another 14% as African American (2023 US Census data).

Nearly all Hispanics have European (and sometimes African) ancestry in addition to their Native American ancestry (5–8), and many African Americans have European (and sometimes Native American) ancestry in addition to their African ancestry (4–5, 8–9). The ancestry of the Brazilian population is even more mixed (10–11). In addition, the fraction of reproductive couples who have widely differing ancestry is steadily increasing (4, 12). Therefore, in the future US there will be many fewer people with ancestry from only one continent, and many more with ancestry from multiple continents. In nations with fewer immigrants, the mixing may be slower, but is still occurring. I don't claim that such mixing will eliminate all racism, but I think there is good reason to hope it will be tempered.

Elitism

It's been said many times that one of the greatest injustices in life is that you don't get to pick your relatives. But I say that an even greater injustice is that you don't get to pick your genome. No one has any appreciable control over their own genome; it is just something with which you are born.

I define elitism (also sometimes called ableism) as discrimination against the less fortunate, including the sick, disabled, and less talented, at the hands of the more fortunate. Elitism is certainly a risk in any program of gene pool management. I think that one of the primary reasons people dislike the subject of gene pool management is that they feel it is just another way in which the more fortunate can take advantage of the less fortunate. I reject any gene pool management program that is elitist.

The leaders of the Eugenics Movement did not treat less fortunate people with respect (Table 15.2). Instead, they used derogatory names such as "imbeciles" and "degenerates." Over my lifetime, I have seen much improvement in the language used by geneticists and others. For instance, people are no longer called mutants, but rather, are said to carry pathogenic variants. I applaud this important change in language. We need to continue to work to ensure that our language conveys respect for all people.

I tried hard in this book to use terms that are not offensive. Nevertheless, I feel that euphemistic language can be carried too far. One important exception to my efforts to avoid offensive language was my use of the word "deleterious," as in "deleterious variants." I carefully considered using alternatives such as "dysfunctional" or "disruptive" or "pathogenic" (which is the term geneticists usually use), but I stuck with "deleterious" to make a point. Although I firmly oppose discrimination against those who are sick or disabled (13), I do not think humanity is better off if, for example, there are more individuals with kidney disease in the world. I believe it takes discipline and perhaps also some strength and courage to understand and accept that while humanity is not better off with more deleterious variants in the gene pool, it is still wrong to denigrate someone because of their genome.

Although our fate is not fixed by our genomes, it would also be a big mistake to discount the role our genomes play in our lives. We all know that life is not fair, but I think we often don't grasp how truly unfair it really is. In terms of our genomes, people are most decidedly *not* equal. Some unlucky people are born with genomes that place them at a significant disadvantage in life. Others are much more fortunate. Many successful people like to explain their success as a result of hard work. While I agree that most successful people do work hard, I also think that many talented people underestimate the *enormous* advantage that they have in life. Many others simply do not have the ability to achieve what they have achieved.

Many presume that if gene management programs were implemented, those with lower reproductive promise would be worse off than they are today. I disagree. To paraphrase lines from the musical *Fiddler on the Roof*, there is no shame in being born with a crummy genome, but it's no great honor either. Many of the least intelligent in our societies live their entire lives in poverty. They are trapped in low-income jobs. They will commit and be the victims of most crime. Through clever advertisements, they will be encouraged to abuse cigarettes and alcohol and to gamble. They will live fewer years. They will be tricked by demagogue politicians. They will be economically dominated and exploited their entire lives by those who are more talented. I see promise in improving the human gene pool and boosting the average intelligence of human populations. If sufficient numbers of people are capable of standing up for themselves, then I think we will see progress on fundamental problems like economic inequality.

Other Risks

Mistakes in gene pool management programs could be costly to our species. Inaction will likely be disastrous, but there is also risk in rash action. I think much additional study of the gene pool is needed. Management programs should only be implemented after careful deliberation. We need to proceed slowly and cautiously.

A majority of people would probably agree that a boost in average intelligence would be beneficial to human societies. But intelligence is only one of many desirable traits. No one knows what combination of traits and talents are optimal for a human society. Any gene management program carries the inherent risk that efforts to boost some traits may diminish other desirable traits. I think we need to retain diversity in human traits.

One approach to minimizing the impact of mistakes in gene pool management is to restrict the size of management programs. Rather than one universal, worldwide gene pool management program, for example, we need to have many independent programs. Each program should have its own criteria and approaches. In this way, mistakes in any one program will not doom our entire species.

I listed as another fatal flaw of the Eugenics Movement that the leaders had a faulty understanding of human genetics (Table 15.2). Geneticists have certainly learned a great deal in the century or so since the height of the Movement, but there is also much that geneticists still do not understand. Is current knowledge sufficient to begin managing the gene pool? I think the answer is probably yes, but I also feel that we need to aggressively pursue additional human genetics research.

Nearly everyone today would probably agree that involuntary sterilization was one of the greatest evils of the Eugenics Movement. I doubt anyone likes the idea of a government telling them whether or not they can have children. Reproductive incentives and/or disincentives may need to be applied, but reproductive decisions must always remain solely with individuals. I therefore reject any gene pool management program that restricts reproductive freedom.

I also reject any gene pool management program that is ordered by a totalitarian government. Gene pool management programs should never be subject to the whims of dictators. The people, through their freely elected leaders, must make the decisions.

Restrictions on Gene Pool Management

I think that risks from gene pool management programs can be reduced by scrupulously abiding by the restrictions listed in Table 16.1.

Table 16.1
Steps to reduce the risks of gene pool management

- Go slowly and act only after careful study and deliberation.
- Reject any gene pool management program that curtails reproductive freedom.
- Reject any gene pool management program that is racist.
- Reject any gene pool management program that diminishes the rights and opportunities of those who are less healthy and talented.
- Retain broad diversity in traits. Programs that aggressively select for just one specific trait like intelligence should be avoided.
- Avoid any gene pool management program that encompasses hundreds of millions or billions of people. Instead, require that nations and regions adopt separate and different gene pool management programs.
- Censure and defeat any gene pool management programs that arise in a totalitarian state. Only programs developed through the slow, messy process of democracy should be employed.

Summary

- Management of the gene pool involves risks to our species.
- These risks can be mitigated by rejecting any gene pool management program that includes any of the following features (see also Table 16.1):
    - Racism
    - Elitism
    - Restrictions on reproductive freedom
    - Rash, aggressive actions
    - Substantial loss of genetic diversity
    - Selection for only one or a few traits
    - Large programs involving hundreds of millions of people
    - Programs developed in totalitarian states

Chapter 17

POSSIBLE SOLUTIONS TO THE GENE POOL PROBLEM

Because of the relaxation of natural selection and other relatively recent changes to evolutionary forces, we can no longer rely on nature to maintain the human gene pool. Deliberate actions on the part of individuals and probably also governments will be necessary. In this chapter, I present some possible solutions to the gene pool problem, emphasizing those solutions that I feel are likely to be the most acceptable and effective.

Overcoming Taboos

If there is only one outcome from Part II of this book, then I hope it will be to stimulate contemplation, discussion, and investigation of the gene pool. The Eugenics Movement and especially the Nazi atrocities swung the pendulum of public (and scientific) opinion strongly against even consideration of the gene pool. It's time to allow the pendulum to swing more freely. The Eugenics Movement was a dark episode in human history that hopefully will never be repeated. But it would be a huge mistake to let it paralyze us forever from attending to the gene pool.

I will go further—I think it is irresponsible to continue to avoid this topic. It is simply too important to ignore. The future of human civilization is at stake. I call on courageous individuals to delve into the gene pool problem. We need to make it an acceptable and frequent subject of discussion at universities, in the media, and throughout our societies.

I plead for people to avoid the word "eugenics" in discussions of the gene pool. This word has such negative connotations that it is impossible to hold productive discussions when it is employed. I recommend that this word be used only to refer to the Eugenics Movement of history. Of course, people who don't like even discussion of the gene pool will label me and others as eugenicists, but I say this is intellectually deceitful. The gene pool problem and its solutions need to be considered dispassionately.

Study of the Gene Pool

My undoubtedly imperfect analysis of the gene pool problem needs to be bolstered by much additional scientific investigation. Knowledge of the gene pool is puny. Some of my thoughts on the most important areas for study are listed in Table 17.1 and discussed below. I raise many questions that need answers. I don't claim that this list is even close to complete, but it should at least offer a few starting points.

Table 17.1
Important areas for gene pool study

- Numbers, types, and impact of mutations
- Possible changes to mutation rates
- Natural history of deleterious DNA variants
- Genealogic demography
- All other evolutionary forces
- Practical and humane solutions to the gene pool problem
- Ideal combinations of traits in future human populations
- Genetics of intelligence

In Table 3.3, I listed the average numbers and types of germline mutations that occur in each newborn. These numbers, however, are just a snapshot of current knowledge; they need to be investigated further and solidified.

Knowledge of the impacts of germline mutations needs to be improved. What fractions of mutations are deleterious, neutral, and advantageous? What is the distribution of penetrance values for deleterious mutations? Do these values depend upon mutation type or location? Are mutations in noncoding regions of the genome more likely to be advantageous than mutations in coding regions?

Geneticists need to determine whether the germline mutation rate is changing. Even small changes in the mutation rate, when compounded over many generations, could have a strong effect on human evolution. Studies completed to date probably rule out large, recent changes to the germline rate (1–4), but small changes due to synthetic chemicals or other agents cannot be excluded (5).

The natural history of deleterious DNA variants refers to the number of generations such variants persist in human populations and how their frequencies wax and wane. How does the natural history vary with penetrance of the variants? Are deleterious variants persisting in populations today for more or less time than in the past? Although geneticists know much less about advantageous variants, the same questions should be posed for them. Since variants that have arisen in a person through mutation can now be easily identified, it will be relatively straightforward to follow them through subsequent generations. Living five-generation families are available today.

To advance the study of genealogical demography, I think that accurate biological family trees for entire populations need to be constructed. These trees will benefit people through both research and directly improved healthcare (see, for example, Case 7 in Chapter 7). With such family trees, it will be possible to determine which individuals make contributions to future gene pools and the range of contributions. Much more also needs to be learned about the average traits of people who have especially large numbers of children.

More investigation is needed into each of the other evolutionary forces that I outlined in Chapters 13 and 14. Is the *direction* of impact on the gene pool correct for recent changes in each of the forces shown in Table 14.4? What is the *relative* impact of each of these forces? Can recent changes in these forces be quantified? Are there other important evolutionary forces that I have not identified?

Table 17.2
Some unanswered questions about the genetics of intelligence

- What is the distribution of effect sizes for DNA variants that adversely affect intelligence?
- What are the DNA features of people with exceptionally high intelligence? Human societies may depend upon geniuses more than we realize.
- Are people with especially high intelligence more susceptible to mental illness or other health problems?
- Over what range do full siblings differ in intelligence, and how can these differences be explained?
- Do parents with below-average intelligence occasionally have a child with significantly above-average intelligence, and if so, is there a genetic explanation?
- Do somatic mutations have a significant effect on intelligence?
- How far can polygenic indices for intelligence be taken?

There are drawbacks to all solutions to the gene pool problem that I have considered. We need many people around the world to devise superior solutions. We need solutions that not only are effective in solving the problem, but also are acceptable to people and don't violate the restrictions shown in Table 16.1.

If governments attempt to incentivize the reproduction of those who are more fortunate in their genomes and/or disincentivize those who are less fortunate, then criteria will need to be developed to categorize those individuals. Should intelligence, health, athleticism, fertility, or other traits be prioritized, and if so, to what relative degree? What mix of traits is needed to allow a human society to thrive? Can analysis of genomes alone determine optimal reproductive potential?

I think that we need to dramatically increase the number of research dollars that are devoted to understanding the genetics of intelligence. It is known that intelligence is heritable, and that deleterious variants in many different genes can cause cognitive deficits, but beyond this, our ignorance is great. There are *many* unanswered questions about the genetics of intelligence; a short list is shown in Table 17.2. In my opinion, we need to devote at least as many resources to the study of the genetics of intelligence as we now devote to, say, the study of diabetes.

Monitoring the Gene Pool

The British engineer William Thomson (1824–1907), who made major contributions to the understanding of electricity and thermodynamics, famously said:

> I often say that when you can measure what you are speaking about, and express it in numbers, you know something about it; but when you cannot express it in numbers, your knowledge is of a meager and unsatisfactory kind; it may be the beginning of knowledge, but you have scarcely, in your thoughts, advanced to the stage of science, whatever the matter may be.

I think Thomson's comments apply perfectly to the gene pool. Only when we can accurately monitor the gene pool will we be certain whether the gene pool is improving or degrading, and only then, will we learn the true rate of change. In the following paragraphs, I describe a few possible approaches to monitoring the gene pool. These approaches can be binned into two groups: those that rely on people's genomes and those that rely on people's traits. Ultimately, some combination of the two may be required. I'll start with approaches that utilize genomes.

In principle, the gene pool could be monitored by simply counting all of the deleterious variants in individuals. A problem with this approach is that although many human genomes have now been sequenced, they haven't been generated from a representative sampling of human populations. Eventually, when genome sequencing becomes near universal in healthcare, the necessary representative data should become available, but we aren't there yet. Another problem is that geneticists haven't yet identified all deleterious variants. A few million deleterious variants are now known (6), but many millions more remain to be characterized. Geneticists are particularly lacking in knowledge of deleterious variants with intermediate penetrance values.

Another approach using genome sequences would be to apply polygenic indices to estimate various traits. Additional investment into research on the genetics of intelligence, as described above, will accelerate improvement in the indices for this trait. When sufficient numbers of representative genomes have been accumulated, and when the indices have sufficiently improved, it should become possible to determine how these traits are changing over time. This approach has already been attempted for educational attainment in Iceland (7).

Improvements in polygenic indices for intelligence and other traits also provide the opportunity to attempt to measure traits in long-deceased individuals through the study of ancient DNA (8). This approach could give geneticists a much longer time frame over which to measure changes to the gene pool. The big problem with this approach is that DNA sequences are available from only a very small fraction of people who lived in the past, and it is unlikely that this small sampling is representative. Still, such studies would be interesting, and possibly enlightening.

As an alternative to using genomes to monitor the gene pool, people's traits can be measured directly (9). Some traits, like height or blood glucose, are relatively easy to measure accurately, but as described in Chapter 12, there is no easy and accurate method for measuring intelligence. A further, perhaps insurmountable, difficulty would be to convince entire populations to submit to intelligence testing.

Probably a better way to use traits to monitor the gene pool would be to more carefully measure differential reproduction. As described below, gene pool management is in essence a matter of differential reproduction. Despite the limitations in available measures of intelligence, it should be possible to accurately determine reproductive rates as a function of these measures, especially as a function of educational attainment.

Reproductive Planning

Many people strive to make the world a better place for succeeding generations. Most parents make great efforts to care for their children, and even to provide for them after death through life insurance and wills. Yet, despite these massive efforts, few people currently take responsibility for one of the actions that will most affect future generations—namely, reproduction.

In the past, reproduction was almost entirely a roll of the dice. Parents had no idea whether their children would be sick or sound. But today, with the accumulated new knowledge and technology described in Chapter 11, it is possible in many cases to avoid single-gene disorders in children. Parents can take responsibility for the genetic health of their children.

In my opinion, by far the best method of avoiding severe disease and disability in children is genetic selection of reproductive partners. People can practice reproductive planning after they have paired, or even after the woman becomes pregnant, but then more difficult decisions about whether to split up or whether to have an abortion often need to be made. Although many may find this extreme, I don't think it makes any sense these days for a knowledgeable person who wants to reproduce responsibly to choose a reproductive partner without knowing whether the partner is genetically compatible. For anyone who is planning to reproduce, getting your genome (and also the genome of your prospective partner) sequenced is an obvious and extremely valuable step.

Differential Reproduction

Studying and monitoring the gene pool and improving reproductive planning will all help to resolve the gene pool problem, but barring the application of the "exotic" approaches discussed below, differential reproduction is the only effective means that I have identified to effectively manage the gene pool. To control the number of deleterious variants in the gene pool, those who are fortunate to be healthier and more talented must out-reproduce those who are less fortunate.

There are some crucial unknowns regarding differential reproduction. One unknown is the magnitude of the difference in reproduction necessary to maintain the gene pool. Just considering intelligence, do those who are more intelligent need to have 1 more child on average than those who are less intelligent, or should the difference be, say, 2.17 children? The difference must be great enough to compensate for mutation, relaxed natural selection, older age at reproduction, other changes to evolutionary forces, and ideally also to make up for the gene pool deficit accumulated throughout the 20th century and continuing today.

Another great unknown is which portions of the distribution should out-reproduce which other portions. Again, for intelligence, should the top 75% out-reproduce the bottom 25%, or should the top 50% out-reproduce the bottom 50%? Obviously, most people need to reproduce to maintain our societies. Unless they had exceptionally large families, it would be totally unworkable, for example, for only the top 25% to reproduce. Investigation of these unknowns will be vital to devising successful solutions to the gene pool problem.

There are some rays of hope for differential reproduction. As described in Chapter 14, the deficit in reproduction among highly educated men is smaller than for highly educated women, and the deficit in highly educated women is beginning to recede in at least some places. If this trend for women continues, then the gene pool problem could conceivably be solved without any intervention by governments. This is another reason why monitoring the gene pool is so important.

Voluntary Actions

Management of the gene pool is obviously a sensitive topic. Mild measures, especially voluntary actions, are likely to be the least controversial, the most widely accepted, and the least expensive for taxpayers. Voluntary measures also maximize individual choice.

For voluntary approaches to have the greatest impact, people must be aware of the gene pool problem. This book will hopefully help a little, but much more is needed. Genetics education needs to be improved at all school levels, from grade school through medical school. Education must continue for adults who are out of school. And, as mentioned above, discussion of the gene pool problem needs to be much more widespread. When large numbers of people understand that their own reproduction impacts the future of our species, then I think we will see at least some improvement in the gene pool problem.

I believe that solving the gene pool problem is of such high importance that I make the following plea. I ask that those who are healthier and more talented have more children than they planned to have before they learned about the gene pool problem. If you would prefer to have no children, please reconsider. If you plan to have two children, please consider having three, four, or more. I particularly direct my request to those who are college educated and to women. I understand this is a huge request. Raising children requires large investments of time and energy. Children often detract from careers and earnings, particularly for women (10–11). But humanity needs more children from those who are genetically fortunate.

A voluntary action that employers can take is to offer generous employee benefits for infant and child care, especially for their highest paid employees. As I former employer, I understand the loss of productivity that results when a valuable employee becomes pregnant, but if generous accommodations are provided to employees with young children, those employees are more likely to be retained. Maternity and paternity leave can seem lengthy to employers, but in reality, it is a relatively short period of time compared to an entire work career.

Limited Government Actions

There are a number of relatively simple and inexpensive steps that governments can take to address the gene pool problem. For a start, governments can encourage and fund genome sequencing for all individuals as a routine part of healthcare. As described in Part I, widespread genome sequencing will significantly improve healthcare. The cost is only a tiny fraction of lifetime healthcare costs.

Governments can also encourage individuals to at least consider genetic reproductive planning. When young people who have had their genomes sequenced at birth reach about 17 years of age, they should be given the option of learning which deleterious variants they carry in their genomes. They should also be given information about reproductive planning, including partner selection. These steps can be mostly completed using computer programs and will therefore incur relatively little cost. Couples who apply to governments for marriage licenses could be *required* to receive genetic counseling and/or DNA testing. These steps were recently implemented in both the United Arab Emirates (12) and Qatar (13). Although it would be more expensive, governments could also consider providing at least some assisted reproductive technologies (Chapter 11) to its citizens free of charge.

Governments can make birth control and elective abortion completely free to all citizens, regardless of ability to pay. In my opinion, no one, especially including those who have been less fortunate in the genome lottery of life, should become pregnant if they want to avoid pregnancy, or should be forced to continue a pregnancy if they want to electively abort. Government measures that encourage reproduction among those who are genetically less fortunate are self-defeating.

More Aggressive Government Actions

I think the moderate government actions described in the previous section should be taken first. However, it may well turn out that more aggressive actions are required. Since I have ruled out infringements on reproductive freedom (Table 16.1), I think this leaves reproductive incentives. There are many possible approaches for such measures; I describe only a few.

One of the primary difficulties with incentives is that they would need to be applied selectively. In other words, individuals would need to be categorized for their reproductive value to societies. To my knowledge, all current approaches to perform such categorization are flawed; some who are deemed to have low reproductive promise would actually have high promise and vice versa. However, the categorization may not need to be perfect to be effective at the societal level. Perhaps the simplest and most effective approach today would be to use educational attainment (at least in developed nations). Educational attainment is relatively easy to measure. Those who do not complete high school are often at a cognitive disadvantage; those who attain a four-year college degree are usually relatively healthy and talented.

Incentives could be offered to individuals with higher educational attainment to encourage them to have more children. For instance, free child care and/or college debt forgiveness could be provided to all *parents* with a four-year college degree. Such incentives could be increased with the number of children and reduced with the parents' age. Free or reduced-cost storage of eggs/sperm for future personal use could also be considered. If individuals who reproduce later in life use sperm and eggs that were collected when they were younger, numbers of mutations will be reduced.

Another possibility is flipping the usual order in life of advanced education and reproduction. For those *parents* with college undergraduate degrees who want to pursue graduate studies, professional internships could be offered with government assistance during the time when they are having babies and raising young children. As an example, young parents who aspire to become physicians could take positions as some type of medical assistant while they are reproducing. This would allow them to continue to learn and gain experience while having babies and help them gain entrance into medical school at a later date. Continued improvements in health and lifespan should allow young people to begin their professional educations a little later in life and still have long, productive professional careers.

Governments could also introduce measures to encourage individuals with low educational attainment to have fewer children. Raising children is difficult for all parents, but especially for parents with limited education and (often) low income. In my opinion, fewer children in families with low education would benefit both the parents and children. It would provide more opportunity for parents to continue their education and develop their careers, and it would allow more time and money to be devoted to each child.

As for possible incentives to encourage individuals with lower education attainment to have fewer children, I think providing free birth control and abortion services, as mentioned above, would be an important step. Child support payments could also be provided that are especially generous for those with one child, but that are reduced per child for those with more than one child.

Incentives are expensive. Tax revenue is always limited and demands on revenue are always great. Therefore, I think it might be most workable to provide incentives primarily to individuals at the extremes of health and talents. This approach would mean that the majority of people with near average health and abilities would be unaffected by the incentives. This could make the measures more politically palatable.

It is difficult for governments to influence the reproductive decisions of their citizens (14). Government leaders in Japan, for example, have been trying to counter reproductive decline for decades without much success (15). Singapore's plan for differential reproduction introduced in 1984 was a total failure (Chapter 15). Incentives definitely have an effect, but they may need to be relatively large to solve the gene pool problem.

Trial and error in many different countries will likely be necessary to find the most effective solutions to the gene pool problem. Persistence will also be an essential component of any effective program. It will likely take at least a few generations for improvement to be apparent even for highly effective programs. Gene pool management is a long-term task for humanity.

Exotic Solutions

I think that differential reproduction will likely be the most effective means of solving the gene pool problem. However, it is possible that other, mostly highly technical solutions will be attempted. A list of some of these exotic solutions is presented in Table 17.3. I briefly discuss each of these possibilities, along with *some* of the obstacles to implementation; all of these approaches face major difficulties.

Population Isolation

It is conceivable that some our healthiest and most talented people could be relocated to one region of the planet or to a space colony. These individuals would then live relatively isolated from the rest of humanity, and could potentially be founders of new, large human populations. A problem with isolated colonies on Earth would be jealousy, friction, and competition with neighbors. The colonies would need to displace most people originally living in their territories and prohibit most immigration. Outsiders would perceive the colonists as conceited elitists. Unless they adopted a very simple, nontechnological lifestyle, the Earth colonies would also be dependent upon outsiders for *many* goods and services. The colonists would also probably need to reproduce at a relatively high rate.

Table 17.3
Exotic approaches for solving the gene pool problem

- Population isolation
- Sperm, egg, and embryo donation
- Embryo selection
- Cloning
- Artificial wombs
- DNA editing
- DNA synthesis

Given that our astronauts have been stringently selected to be among our brightest and best, I think it likely that the first space colonies will be populated by some of our healthiest and most talented. However, problems such as dependence upon Earth and friction with other human groups would apply to the space colonies just as with isolated Earth colonies. In addition, all isolated colonies, whether in space or on Earth, would always need to deal with children who have diminished health and abilities due to mutation. The colonists would presumably need to either expel such individuals or inhibit their reproduction.

Sperm, Egg, and Embryo Donation

Sperm, eggs, or embryos could be donated by healthy, talented individuals and then used to impregnate women through artificial insemination (AI) or in vitro fertilization (IVF). This was the concept behind the Nobel Prize Sperm Bank described in Chapter 15. The basic problem with this approach is that parents prefer to carry, deliver, and raise children who are biologically related to themselves. People who are infertile may be willing to accept donated sperm, eggs, or embryos, but fertile individuals usually are not.

Over about the last two decades, it has become possible to reverse and redirect the development of cells. In nature, early embryonic cells differentiate into all of the various cell types in our bodies, but this differentiation is always one way. Skin cells, for instance, don't naturally convert into muscle or nerve cells. However, scientists have recently learned how to achieve such conversions in cell culture outside the body. Many cell types can be converted into "pluripotent" cells capable of differentiating into nearly all cell types, including muscle and nerve cells (16–17). Recently, these pluripotent cells have also been converted into precursors of egg cells (18–20). Although this technology is still at an early stage, within a relatively few years it may become possible to generate a virtually unlimited number of egg cells from donors, and unlimited numbers of embryos from couples. It is far from clear, however, whether this technology can be safely used to produce healthy children. The relatively high rates of somatic mutation and the considerable manipulation of the cells in culture may appreciably affect the health of offspring.

Embryo Selection

In Chapter 11, I described how cells could be removed from embryos generated by IVF for preimplantation genetic testing. Today, this technology is used primarily to avoid single-gene disorders and aneuploidies. However, there are no technical barriers to performing more extensive DNA testing on the embryos, including full genome sequencing. Genome sequencing allows polygenic risk indices to be generated for an unlimited number of traits and disorders. Parents could then select what they consider to be the genetically best embryo for transfer into the uterus. Some companies are already beginning to offer this service.

There are several problems with embryo selection, including the problem of trade-offs. How, for example, do parents choose between an embryo with a relatively high index for intelligence but also with poorer indices for cancer and diabetes, and another embryo that is likely to result in a healthier child but one with lower intelligence? Another problem is the current crude nature of most risk indices. They don't yet do a great job of predicting traits and disorders (Figure 12.4). Finally, embryo selection is expensive. Without substantial government support, only the wealthy could afford such a process. High cost is a problem with nearly all of the exotic approaches.

Cloning

Clones are individuals, other than identical twins, who are genetically identical (21). Today, cloning is typically accomplished by fusing a somatic cell from a donor with an egg cell that has had its nucleus (along with all DNA except for mitochondrial DNA) removed. The resulting embryo is then transferred into a uterus, and the resulting newborn is a clone of the somatic cell donor. The first mammal to be cloned was the sheep Dolly in 1996. Cloning of dogs and cats is now relatively commonplace (22). Although apparently no humans have yet been cloned, monkeys have (23), and so there are probably no insurmountable technical obstacles to human cloning. In principle, given sufficient money, it should be possible to produce a large number of genetically identical babies from particularly healthy and talented individuals. However, the same objection that people want to raise their own biological children also applies to cloning. In addition, human cloning is generally viewed with repugnance. Every nation that has considered the matter has prohibited the practice (24). Another problem with cloning is that it would eliminate meiosis and the accompanying meiotic recombination that is essential for long-term maintenance of human health and vigor (25).

Artificial Wombs

In an attempt to decrease morbidity and mortality in babies born prematurely, researchers are developing "artificial wombs" (26–27). The current stage of this research is *far* from enabling the complete gestation of a baby from embryo to birth outside of a woman's body, but if this technology continues to advance, then it is conceivable that someday it will be possible to produce babies without pregnancy. Because this would reduce the burden and health risks of reproduction, it might encourage talented women who are devoted to their careers to have more children.

DNA Editing

Over the last few decades, geneticists have developed sophisticated methods for editing DNA sequences in any organism. Application of these methods, through what is often called genetic engineering, has enabled the genomes of many species to be modified to serve human purposes. In humans, DNA editing—usually called gene therapy—has been applied to *somatic* cells in efforts to treat cancer and a rapidly growing list of single-gene disorders (28–29). Editing of somatic cells does not affect the egg and sperm cells, so resulting changes are not passed on to children. In contrast, *germline* DNA editing in humans, which would affect the genomes of children, has to date been prohibited by governments.

In principle, if DNA editing became sufficiently inexpensive, it could be used to modify the genomes of large numbers of people. Some futurists have predicted that DNA editing will be used to radically alter humans, such as making people resistant to all viruses or capable of using sunlight as an energy source, like plants (see, for example, 30–32). One of the many barriers to germline DNA editing is that the required experiments on people would strike many, including me, as unethical. Imagine that, based on studies in mice and monkeys, a group of geneticists becomes convinced that inserting two genes into human genomes would boost intelligence by 5%, but to prove their hypothesis, they would need to edit the genomes of 100 embryos. They estimate that 2 of the treated embryos would lead to children with significant health problems. Would it be ethical to edit the embryos' genomes in this way without the children's consent, knowing that intelligence might not even be boosted as predicted?

DNA Synthesis

Biochemists have learned how to synthesize fragments of DNA using chemicals produced from petroleum. Synthetic DNA fragments in the range of 10–100 nucleotides are now used ubiquitously in genetics research and clinical DNA testing. Longer DNA fragments can also be synthesized and stitched together to produce entire genomes. Genomes of viruses (thousands of nucleotides) and bacteria (one to a few millions of nucleotides) have now been synthesized, and after being introduced into cells, found to be perfectly capable of supporting life (33–34). Geneticists are close to completing the synthetic genome of brewers' yeast (12 million nucleotides) (35–37) and have targeted the genome of a moss (480 million nucleotides) (38). Although the genomes of mammals (about 3 billion nucleotides) are currently out of reach, it is not hard to project that researchers will get there some day (34). With synthetic human genomes, it would be possible to change many sites in the genome at one time. Theoretically, it would be possible to create genetic "super people." Among the many problems with synthetic genomes are the need for extensive, ethically dubious, experimentation and conflicts with "natural" people. The latter objection has been exploited many times by fiction writers.

- - -

Because of all the problems with the exotic solutions, and possibly also because I'm old and uncomfortable with these approaches, I strongly favor differential reproduction as the best solution to the gene pool problem. However, I don't pretend to be able to predict the future. It's certainly possible that one or more of the exotic approaches that I've described (or another) will ultimately prove to be the solution to the gene pool problem. People's aversions to new technologies can and have changed. When AI and IVF were first introduced, most people were opposed, but now most (although certainly not all) accept them without concern.

Summary

- To improve our understanding of the gene pool problem and to devise effective solutions, we need to overcome the taboos now associated with this subject. We need to openly discuss and investigate the gene pool.
- Use of the word "eugenics" should be restricted to the Eugenics Movement of history.
- Much more needs to be learned about the genetics of intelligence.

- Accurate means of measuring and monitoring the gene pool need to be developed.
- Barring exotic approaches, differential reproduction will be necessary to solve the gene pool problem. Just to maintain the status quo, those who are fortunate to be healthier and more talented need to out-reproduce those less fortunate.
- Voluntary measures should at least help to solve the gene pool problem. Governments can take a number of steps to facilitate voluntary approaches.
- If voluntary approaches prove insufficient, then financial incentives or disincentives may be required to increase the reproduction of those more genetically fortunate and/or to decrease the reproduction of those less fortunate. There are many possible options for such incentives/disincentives.
- Several possible exotic solutions to the gene pool problem exist, as listed in Table 17.3, but all of these exotic approaches currently have significant drawbacks.

Chapter 18

SUMMARY OF PART II, THE GENE POOL

Brief Review of Chapters 9–17

In Part II, I described how reproductive planning can be used in many cases to avoid severe health problems and disabilities in children. I outlined the abundant evidence that human intelligence has substantial heritability. I covered forces involved in the evolution of our special intelligence and explored how these forces have changed over about the last 200 years. Through this analysis, I identified the gene pool problem. Finally, I presented some possible approaches to gene pool management—approaches that avoid racism, elitism, and other risks.

The summaries at the ends of each chapter provide a quick means of reviewing the major themes of Part II. I encourage readers to reread these points. To facilitate this step, I have repeated the summary points here. Statements that are my personal opinions rather than scientific facts are marked by asterisks (***).

Chapter 10: Prenatal Development and Prenatal DNA Testing

- Prenatal development is a complex 38-week process that begins with a fertilized egg cell and ends with the birth of an infant.
- Some of the cells from division of the fertilized egg cell develop into the baby and others develop into the extraembryonic tissues: the amnion, chorion, placenta, and umbilical cord.

- Many things can and often do go wrong during prenatal development. Roughly 60% of all pregnancies and 15% of recognized pregnancies spontaneously abort.
- The most frequent known cause of spontaneous abortion is aneuploidy, or an abnormal number of chromosomes. Nearly all aneuploid embryos and fetuses spontaneously abort.
- DNA from the developing fetus may be tested through both invasive and non-invasive procedures.
- Invasive testing is practiced on only a small fraction of pregnancies and nearly always involves either chorionic villus sampling (typically performed at 10–13 weeks of pregnancy) or amniotic fluid collection (typically performed at 15–20 weeks).
- All types of clinical DNA tests may be performed on fetal DNA obtained from the invasive procedures.
- Non-invasive prenatal screening (NIPS) is a maternal blood test used to assay the fetal DNA. Today, NIPS is used primarily to test for aneuploidies, but in the future may be used to test for other genetic abnormalities.
- In developed countries, NIPS for aneuploidy detection is currently performed on a substantial fraction of all pregnancies.

Chapter 11: Reproductive Planning

- Although single-gene disorders are individually rare, because there are thousands of them, total societal costs of caring for people with single-gene disorders are high. These costs include both direct healthcare costs and indirect costs to the caregivers.
- Assisted reproductive technologies (ARTs) have been developed to help subfertile and infertile couples reproduce and avoid genetic disease. ARTs include artificial insemination (AI), in vitro fertilization (IVF), and preimplantation genetic testing (PGT).
- Through DNA testing, geneticists can now predict a reproductive couple's chance of having a child with a single-gene disorder. This ability may expand in future to include genetically complex disorders.

- Couples can avoid the birth of a child with a genetic disorder via several options, including partner selection, prenatal testing, and ARTs. Abortion can be avoided through partner selection and ARTs.
- Preconception reproductive planning through selecting a genetically compatible partner is an attractive option that will hopefully grow in popularity.***
- At least in the near future, genetic reproductive planning will probably have only a limited impact on the gene pool.***

Chapter 12: Genetics of Intelligence

- Intelligence, broadly defined as the ability to solve problems, varies over a wide range among individuals.
- Higher intelligence is a significant factor in both individual and societal success. Human civilizations could not have arisen without our special intelligence.
- There are multiple compelling lines of evidence that intelligence has significant heritability. These include family studies, twin studies, adoption studies, and the genetics of intellectual disability (ID) (Table 12.3).
- Geneticists have now identified the specific DNA causes of many thousands of cases of ID. A change in a single nucleotide out of six billion can spell the difference between a person with normal intelligence and a person with ID.
- Although different studies have produced a range of estimates for the heritability of intelligence, it is a scientific *fact* that intelligence has substantial heritability (on the order of 60%).
- Nonheritable factors also strongly influence intelligence. These include mutations and environmental factors such as premature birth, lead poisoning, and upbringing.
- The mutations that occur in every child in every generation (Table 3.3) cause frank ID in about 1 out of every 100 newborns, and somewhat reduced innate intelligence in an unknown larger fraction of newborns.

- Germline mutation relentlessly erodes the intelligence of human populations.

Chapter 13: Evolution of Our Special Intelligence

- Over about the last seven million years, since the time of the common human-chimpanzee ancestor, humans evolved a special level of intelligence that is unique among all species on the planet.
- Although the process that led to our special intelligence was undoubtedly complex and is still largely a mystery, three *facts* about the process are known:
    - Our special intelligence evolved through changes to our genomes.
    - Mutation created the new DNA variants that increased the intelligence of our ancestors.
    - On average, our more intelligent ancestors reproduced at higher rates than our less intelligent ancestors.
- Geneticists have identified at least several of the major forces that were likely involved in the evolution of our special intelligence (Table 13.2). There is considerable overlap among many of the forces.
- Differential reproduction is defined as differences in reproductive rates among people who carry different DNA variants.
- Natural selection acts by decreasing the frequencies of deleterious variants and increasing the frequencies of advantageous variants.
- Genealogical demography is the study of the numbers and characteristics of people's ancestors and descendants. Many people who lived in the past made no or only small contributions to the current gene pool, and others made large contributions.
- Polygyny, which means one male reproducing together with multiple females, was likely common throughout human evolutionary history.
- Changes in population size can affect the gene pool by altering the level of genetic diversity and the rates at which variant frequencies change.
- Migration leads to new combinations of variants.

- Infanticide, the killing of infants, was in the past common in many, perhaps nearly all, human societies.
- Assortive mating means that women and men who are similar in various traits, such as height or education level, tend to reproduce together.
- Population crashes, defined as relatively rapid drops in population size, were common in our past.
- Random chance events also played a role, perhaps a major role, in evolution of our special intelligence.

Chapter 14: The Problem: Recent Changes in Evolutionary Forces

- In the past, the various evolutionary forces described in Chapter 13 kept mutation at bay by efficiently removing new deleterious variants from the gene pool.
- As a result of relatively recent changes to the evolutionary forces (summarized in Table 14.4), deleterious variants are now accumulating in the gene pool. Average health and abilities of our species are declining. This is the gene pool problem.***
- Many other geneticists have recognized the gene pool problem in modern times, beginning with Francis Galton and Charles Darwin over 100 years ago.
- The rate at which the gene pool is degrading is unknown, but it would be dangerous to ignore this problem.***
- If my analysis is correct, and if we want to maintain our advanced human societies, then we *must* solve the gene pool problem. We have no choice.***

Chapter 15: History of Gene Pool Management

- Gene pool management is not a new concept. Incest avoidance has probably been practiced in the human lineage for millions of years.
- Classical Greek societies considered and implemented gene pool management.

- The Eugenics Movement started in England in the 1880s and for a few decades was popular in a number of countries, including the US. The movement was fatally flawed for a number of reasons (Table 15.2), including racism, poor understanding of genetics, and restriction of reproductive freedom.
- The Nazis in Germany in the 1930s and 1940s carried ideas from the Eugenics Movement to a horrific extreme. The Nazis sterilized about 400,000 people and murdered another approximately 240,000 people they deemed genetically flawed.
- Aggressive gene pool management efforts in Singapore in the 1980s failed.
- Some forms of gene pool management are being practiced today. These include avoidance of consanguineous matings, sperm donation, and reproductive planning.

Chapter 16: Racism, Elitism, and Other Risks

- Management of the gene pool involves risks to our species.***
- These risks can be mitigated by rejecting any gene pool management program that includes any of the following features (see also Table 16.1).***
    - Racism
    - Elitism
    - Restrictions on reproductive freedom
    - Rash, aggressive actions
    - Substantial loss of genetic diversity
    - Selection for only one or a few traits
    - Large programs involving hundreds of millions of people
    - Programs developed in totalitarian states

Chapter 17: Possible Solutions to the Gene Pool Problem

- To improve our understanding of the gene pool problem and to devise effective solutions, we need to overcome the taboos that are now associated with this subject. We need to openly discuss and investigate the gene pool.***

- Use of the word "eugenics" should be restricted to the Eugenics Movement of history.***
- Much more needs to be learned about the genetics of intelligence.***
- Accurate means of measuring and monitoring the gene pool need to be developed.***
- Barring exotic approaches, differential reproduction will be necessary to solve the gene pool problem. Just to maintain the status quo, those who are fortunate to be healthier and more talented need to out-reproduce those less fortunate.
- Voluntary measures should at least help to solve the gene pool problem. Governments can take a number of steps to facilitate voluntary approaches.***
- If voluntary approaches prove insufficient, then financial incentives or disincentives may be required to increase the reproduction of those most genetically fortunate and/or to decrease the reproduction of those least genetically fortunate. There are many possible options for such incentives/disincentives.***
- Several possible exotic solutions to the gene pool problem exist, as listed in Table 17.3, but all of these exotic approaches currently have significant drawbacks.

Those Who Are Genetically Less Fortunate

All people want dignity and respect. Many people also have a strong drive to reproduce. In my opinion, gene pool management should never be used as a club to bash those less fortunate in the genetic lottery of life. Listed in Table 18.1 are some of my thoughts about those who are genetically less fortunate.

Table 18.1

Thoughts about people who are genetically less fortunate

- My respect for people does not depend upon their genome, but rather upon the decisions that they make and the actions that they take during their lives. Many who are genetically fortunate live deplorable lives, and many who are not live sterling lives.
- No one can justly be blamed for their genome. No one should ever be ashamed of their genome.
- There is no such thing as a genetically perfect person. Everyone is genetically flawed.
- All people should be entitled to the same human rights and opportunities, regardless of their genomes. No one should ever accept the idea that some are more entitled because they are genetically more fortunate.
- All current approaches for measuring reproductive promise are flawed. A person may have higher or lower reproductive promise than indicated.
- Everyone should have the freedom to have as many children as they like.
- Some people require more support than others. That does not make them selfish.
- Gene pool management is not punishment.
- We all have a duty toward others. Part of that duty involves planning our reproduction to promote the overall good of society.
- Having only a single child may benefit both the parent and the child. For the parent, it may provide a better chance to advance their education and career. For the child, it may improve their home environment.
- We all have the right to laughter, love, happiness, and a wonderful life.

Final Words on the Gene Pool Problem

It's important to understand that the gene pool problem is not based on racism or elitism, but rather on science. I think nearly all people, myself included, wish that we didn't have to worry about the gene pool, but my scientific analysis indicates that deleterious DNA variants are accumulating and that this situation needs to be addressed. Although our special intelligence distinguishes us from all other species on Earth, humans are not immune from the laws of genetics.

Heritabilities of nearly all human traits, including intelligence, are substantially less than 100%. Average human intelligence could be boosted solely by improving the environmental factors that influence intelligence. Why not, therefore, just focus on the environment and forget about the gene pool? The reason is that environmental improvements in the presence of a deteriorating gene pool will only be a short-term fix. No matter how hard we try, if the gene pool is deteriorating, then efforts to improve the environment will ultimately fail. This does not mean, however, that environmental factors are unimportant or that they should be ignored. The environment is probably more malleable than the gene pool.

When considering the gene pool, we need to broaden our time horizons. People are experienced at dealing with time frames of months, years, or occasionally a few decades, but we rarely consider time frames of centuries or millennia. Yet this is just what is needed for the gene pool. Persistence will be an essential component of any successful gene pool management program.

Experts have compiled lists of threats that could result in the extinction or irreversible decline of our species (see, for example, 1–4). I think the gene pool problem needs to be added to these lists. If our civilizations crumble because of degradation of the gene pool, then although our species may survive, we will have lost what it really means to be human. Maybe if such a calamity occurred, our species could recover and regain civilization, but there is no guarantee. As I pointed out at the end of Chapter 13, evolution of higher intelligence on our planet has been an exceptionally rare event. It may not happen again.

GLOSSARY

GLOSSARY

Advantageous mutation/variant: A mutation or variant that enhances the health or abilities of an individual.

Amniocentesis: An invasive prenatal test, usually performed between 15 and 20 weeks of pregnancy, in which a small volume of the amniotic fluid surrounding the fetus is collected. The fluid contains fetal cells that may be used for DNA testing.

Aneuploidy: An abnormal number of chromosomes in an individual. The only human chromosomal abnormalities that are compatible with life are trisomies 13, 18 and 21, and variations in the numbers of the sex chromosomes.

Artificial insemination (AI): The artificial introduction of sperm from a donor into the uterus of a woman who intends to become pregnant.

Artificial selection: The intentional human selection for desirable traits through controlled breeding of a species. Farmers have used artificial selection for millennia to improve yields in agriculturally important plant species (Figure 3.9). Breeders of animals such as dogs also used artificial selection to achieve remarkable variation within species (Figure 3.10).

Assisted reproductive technologies (ARTs): Various, mostly newer, technologies that were developed to help infertile couples have children. ARTs include artificial insemination, in vitro fertilization, and preimplantation genetic testing.

Assortive mating: People choosing reproductive partners who are similar to themselves for various traits.

Autosomal chromosomes (or Autosomes): The numbered chromosomes 1-22. Chromosome 1 is the largest, and chromosome 21 is the smallest. Compare with sex chromosomes.

Base: A component of a nucleotide. There are four bases in both DNA and RNA (Figures 2.5 and 2.6). In DNA, an A base always pairs with a T base, and a C base always pairs with a G base.

Carrier: A person who is heterozygous or homozygous for a DNA variant of interest. If the variant is rare, the frequency of homozygotes will be very low, and therefore carrier often refers just to heterozygotes.

Centromere: The place along a replicated chromosome where the chromosome arms are bound together. Centromeres appear in the microscope as constrictions in the chromosome arms.

Chorionic villus sampling (CVS): An invasive prenatal test, usually performed between 10 and 13 weeks of pregnancy, in which a small specimen of the chorionic villi (extraembryonic tissue) is collected. Chorionic villi have the same DNA sequences as the fetus.

Chromosome: A highly compacted macromolecular complex consisting of one very long, thin DNA molecule and *many* protein molecules.

Clinical feature (also called symptom or phenotype): A trait or laboratory result outside of the normal range. Examples are abnormal convolutions of the brain surface (lissencephaly), eyes that are farther apart than normal (hypertelorism), unusually fragile bones (osteopenia), slower than normal infant and child development (developmental delay), and high blood sugar (hyperglycemia). All disorders, whether single gene or complex, are characterized by their clinical features (Table 4.5).

Complex disorders and traits: Health problems or traits that are caused and influenced by variants in multiple genes and usually also by non-heritable factors.

Concordance: The degree of similarity in a specific trait between a pair of individuals or two groups of individuals.

Consanguineous matings: Reproduction between relatively closely related individuals such as cousins. Consanguineous matings include incestuous matings, which usually refer to parent-child or sibling matings, but is a broader term that includes reproduction among more distantly related individuals.

Cytoplasm: The outer portion of a cell. Protein manufacture takes place in the cytoplasm.

Disability: A physical or mental condition that limits a person's movement, senses or activities. Disease and disability often, but do not always, overlap. For example, clinical features of trisomy 21 include intellectual disability, but also heart disease and susceptibility to leukemia.

Deleterious mutation/variant (also called pathogenic mutation/variant): A mutation or variant that diminishes the health or abilities of an individual. Some deleterious variants only manifest their effect when present in both copies of a gene.

Disease (or Disorder): An abnormal condition resulting in pain, dysfunction, distress, weakness or death. A disease is a significant deviation from normal healthy life. Diseases usually result from a malfunction or disruption of one or more biological systems, and are described by a set of clinical features.

DNA: Deoxyribonucleic acid. DNA is the genetic material. It is a linear, unbranched polymer composed of four chemical units called nucleotides (Figures 2.5 and 2.6).

DNA banking: The long-term, secure storage of a person's DNA, even after the person's death, for the purpose of future testing.

DNA sequence variant: See Variant.

Dominant (or autosomal dominant) single gene disorder: A single gene disorder in which affected individuals have one dysfunctional copy and one functional copy of a gene located on one of the autosomes. The dysfunctional gene can be inherited from either of the parents, or can arise through mutation. When inherited, each child has an independent, 50% chance of receiving the dysfunctional gene (Figure 4.4B). Contrast with recessive and X-linked single gene disorders.

Egg cell (also called ovum): The female cell that can be fertilized by a sperm cell and that develops into a baby. All people arise by cell division from a single fertilized egg cell.

Elective abortion: Termination of a pregnancy that occurs through a deliberate, intentional procedure. Contrast with spontaneous abortion.

Embryo: A developing baby during about the first 8 weeks of pregnancy post fertilization.

Enzyme: A protein that catalyzes (speeds) a specific chemical reaction.

Exome: All of the exons within a genome. The human exome comprises about 1.5% of the total genome.

Exome sequencing: Sequencing of just the exons of the genes. Exome sequencing covers roughly 1.5% of the total nucleotides in the genome. Contrast with genome sequencing (Table 7.4).

Exon: A segment of a gene that is present in the mature mRNA (Figure 2.19).

Family tree drawing (also called kindred or pedigree drawing): A drawing that indicates the biological relationships among multiple generations of family members. By convention in family tree drawings, each generation is listed on a separate row, males are indicated by squares, and females by circles (an example in shown in Figure 4.7).

Fertilization (also called conception): The fusion of a sperm and egg cell leading to the next generation.

Fetus: A developing baby after about the first 8 weeks of pregnancy post fertilization until birth.

Fixed variant: A new variant that achieves 100% frequency in a population over many generations.

Founder variants: Deleterious variants, usually causative for recessive disease, that have relatively high frequencies in reproductively isolated populations (Table 13.4).

Frameshift mutation/variant: A mutation or variant that, through the loss or addition of a number of nucleotides other than integer multiples of three, changes the reading frame of translation. Frameshift variants usually create new premature stop codes downstream of the site of deletion or insertion (Figure 3.5). Frameshift variants can also extend the length of proteins by eliminating a normal stop code.

Fraternal twins (also called dizygotic or two egg twins): Twins that arise from the fertilization of two separate egg cells by two separate sperm cells (Figure 4.5). Fraternal twins share 50% of their DNA, and are no more genetically alike than any pair of full siblings. Fraternal twins may be same sex or opposite sex.

Gene: The segment of a DNA molecule that encodes a single protein.

Gene pool: All the DNA molecules within people currently living. Gene pool can refer to all people living anywhere on the planet (the usual meaning in this book), or to a smaller population, such as the gene pool of Switzerland.

Gene pool management: Active steps taken to prevent single gene disorders and to control the number of deleterious variants in the gene pool.

Gene regulation: See Regulation of genes.

Gene therapy: A treatment for single gene disorders that involves overcoming a dysfunctional gene by adding a normal gene, silencing the dysfunctional gene, or directly correcting the dysfunctional gene. Several gene therapy treatments have already been approved for use in patients, and many others are under development.

Genealogical demography: A branch of demography concerned with the numbers and characteristics of ancestors and descendants of individuals.

Genetic engineering: The deliberate modification of the genomes of individuals of a species to serve human purposes. Many economically important plant and animal species have already been genetically engineered. Genetic engineering of somatic cells (but not germ cells) of people (usually called gene therapy) is now being performed to treat single gene disorders.

Genome: The total complement of DNA molecules within a species or individual.

Genome sequencing (also called full or whole genome sequencing): Sequencing of the entire genome of an individual. Contrast with exome sequencing (Table 7.4).

Geoancestry: The continental origin of a person's ancestors as of about the year 1500.

Germ cells: The egg cells within the ovaries in females and the sperm cells within the testes in males.

Germline: The lineage of cells leading to the sperm and egg cells. Germline DNA testing means testing of the DNA molecules that are present in all of a person's cells. Germline mutations are mutations that occur in the egg and sperm cells or the cellular precursors to the egg and sperm.

Healthcare provider: A professional provider of healthcare services. Healthcare providers can be physicians, genetic counselors, nurses, physician assistants, psychologists, etc.

Hemoglobin: The protein in blood that ferries oxygen from the lungs to all the cells and returns carbon dioxide from the cells to the lungs. Hemoglobin is a complex composed of four protein subunits (two alpha and two beta protein molecules) along with a non-protein iron-containing molecule called heme.

Heritability: The fraction of variability in a disorder or trait that is attributable to variations in inherited DNA. Heritability values range from 0% to 100%.

Heterosis: The phenomenon, mostly known from plant breeding, in which the offspring of crosses between inbred parental strains show greater vigor than either parent. There is evidence that heterosis occurs in humans through matings between people who are not related.

Heterozygous: Having two different nucleotides at a specific position in a pair of chromosomal DNA molecules of one type (Figure 4.3). People can be heterozygous for nucleotide substitutions, deletions or insertions, or other types of variants.

Homozygous: Having the same nucleotide at a specific position in a pair of chromosomal DNA molecules of one type (Figure 4.3). People can be homozygous for nucleotide substitutions, deletions or insertions, or other types of variants.

Human genetics: The science of the causes and the inheritance of human variation.

Human reference genome sequence: The genomic sequence with the most frequent (the consensus) nucleotide among unrelated individuals at each position along each DNA molecule. DNA sequence variants are typically detected as differences between the DNA sequences of individuals and the reference sequence (Figure 5.8). The human reference sequence is freely available through the internet, and is infrequently updated.

Identical twins (also called monozygotic or one egg twins): Twins that arise from the splitting of an early embryo derived from a single fertilized egg cell into two clumps of cells, each of which develops into a baby (Figure 4.5). Identical twins are always same sex and are 100% alike in their DNA sequences (except for somatic mutations).

Inbred: Plants or lab animals that have undergone many generations of self or brother-sister matings until essentially all DNA diversity has been eliminated. Except for mutations, each individual in each inbred strain is homozygous for the same nucleotide at each chromosomal DNA position.

Inbreeding: Mating between related individuals.

Incest: Reproduction among closely related individuals, usually parent-child or sibling matings.

Infanticide: The killing of infants, usually by their parents, either directly or through abandonment (often called exposure).

Intellectual disability (ID): Intelligence within the lowest 2.5% of test scores.

Intracytoplasmic sperm injection (ICSI): The in vitro fertilization of an egg cell by injection of a single sperm cell into an egg cell.

Intron: A segment of a gene between exons (Figure 2.19).

Invasive prenatal testing: Prenatal testing in which a small tissue specimen, nearly always chorionic villi or amniotic fluid, is collected from the fetus for testing.

In Vitro: Outside the body. Refers to laboratory experiments and procedures that take place outside of living organisms.

In Vitro fertilization (IVF): Fertilization of an egg cell that takes place in a plastic container outside of the body.

Loss-of-Function mutation/variant: A mutation or variant that results in complete loss of the function of the coded protein. Loss-of-function variants include those that delete an entire gene or a significant part of a gene, nucleotide substitutions that create premature stop codes (nonsense variants), deletions or insertions of numbers of nucleotides other than integer multiples of three that change the reading frame of translation (frameshift variants) (Figure 3.5), and variants that disrupt normal intron splicing.

Macromolecule: A large molecule. Examples include DNA, RNA, and proteins. Nearly all of the macromolecules found in living organisms are polymers.

Meiosis: The process of chromosome distribution that occurs during the formation of the egg and sperm cells (Figures 2.16 and 4.1). Both egg and sperm cells have only 23 chromosomes, one of each type.

Messenger RNA (mRNA): An RNA molecule complementary in sequence to a DNA template that is used to manufacture a protein. mRNA is the intermediary that conveys the information within a sequence of nucleotides along a DNA molecule into the sequence of amino acids in a protein.

Miscarriage: A spontaneous abortion that occurs during the first half of a normal pregnancy (less than about 20 weeks).

Missense mutation/variant: A mutation or variant within an exon that results in an amino acid substitution in the coded protein. Such mutations/variants often, although not always, alter the function of the protein.

Mitochondria: Subcellular organelles that generate chemical energy for cells. Mitochondria are the cell's power plants. Each mitochondrion has a relatively small (approximately 16,500 nucleotide) chromosome.

The mitochondrial chromosome is present in egg cells but not in sperm cells, and is therefore inherited only from mothers.

Mitosis: The process of chromosome distribution that occurs during the division of somatic cells (Figure 2.15). The result of mitosis is two daughter cells, each with 46 chromosomes.

Monogamy: People reproducing during their lives with only a single partner.

Monosomy: One copy of a specific chromosome in an individual. An example is one copy of the X chromosome in women with Turner syndrome.

Mutation: A change in the nucleotide sequence of a DNA molecule. Mutations result in DNA sequence variants. Mutation is a normal and ubiquitous biological process. Mutations are deleterious, neutral or advantageous, and vary in the strength of their effect. Mutations can also be either germline or somatic.

Natural selection: The process by which nature selects for individuals with the greatest ability to survive and reproduce in their particular environment. Natural selection leads to decreases in population frequencies of deleterious variants and increases in frequencies of advantageous variants.

Negative selection (or negative natural selection): An evolutionary process that results in a decrease in population frequency of deleterious variants because individuals with these variants produce fewer descendants than individuals without these variants.

Neutral mutation/variant (also called benign mutation/variant): A mutation or variant that does not appreciably affect the health or abilities of an individual.

Non-invasive prenatal screening (NIPS): The testing of cell-free fetal DNA present in the mother's blood during pregnancy. NIPS is also often called non-invasive prenatal testing (NIPT).

Nucleotide: The fundamental molecular unit (monomer) of DNA and RNA polymers. Each nucleotide consists of a phosphate, a sugar (deoxyribose in DNA and ribose in RNA), and a base (Figure 2.5). Lengths of DNA segments are measured in nucleotides, although the word "base" is often substituted for nucleotide. For example, kb is an abbreviation for thousands of bases/nucleotides, and mb is an abbreviation for millions of bases/nucleotides.

Nucleus: A membrane-enclosed organelle within the cell that contains all the chromosomes except for the small mitochondrial chromosome. Replication, production of RNA molecules, and splicing all take place within the nucleus.

Outbred: Lab animals or plants that have not been inbred, and that retain substantial DNA diversity.

Penetrance: A property of variants defined as the fraction of people with one or more specific variants who are affected with a disorder or have a particular trait. Penetrance values range from 0% to 100%. For many variants involved in many different disorders and traits, penetrance is incomplete (less than 100%).

Pharmacogenetics (or pharmacogenomics): The use of genetics to improve the efficacy and safety of prescription drugs.

Phase: In genetics, phase means whether two or more sequence variants are present on only one or both chromosomes of a type (Figure 7.2). Relatedly, it is sometimes also important to determine whether the variants are located on the maternally or paternally inherited chromosome.

Phenotype: See Clinical feature.

Placenta: The organ that serves as the intermediary between the mother and fetus during pregnancy. The placenta develops from embryonic cells and therefore has the same DNA sequences as the fetus.

Polyandry: One woman reproducing with multiple men.

Polygenic index (also called polygenic score or polygenic risk score): A score determined by combining many individual DNA risk factors to predict the likelihood that a person will develop a specific genetically complex disorder or trait. Contrast with risk index.

Polygyny: One man reproducing with multiple women.

Polymer: A chain-like macromolecule composed of repeating chemical units joined together.

Population bottleneck: A temporary (anywhere from a few to many generations) reduction in population size.

Population genetics: The mathematical branch of genetics dealing with changes in the frequencies of variants in populations.

Positive selection (or positive natural selection): An evolutionary process that results in an increase in population frequency of advantageous variants because individuals with these variants produce more descendants than individuals without these variants.

Precision (also called personalized or individualized) medicine: The treatment and management of patients not on the basis of their broad disease group, such as all cancer patients or all epilepsy patients, but rather on the basis of their individual DNA sequences or other individual features.

Preconception reproductive planning: Reproductive planning that begins before conception. Choosing a reproductive partner based (in part) on the person's genome is one form of preconception reproductive planning.

Preimplantation genetic testing (PGT) (also called preimplantation genetic diagnosis (PGD)): DNA testing of an embryo resulting from IVF prior to transfer of the embryo into the mother's uterus.

Prenatal DNA testing: Testing of DNA from a fetus before birth. Prenatal testing can be invasive, in which a small tissue specimen from the fetus is collected and tested, or non-invasive, in which fetal cell-free DNA from the mother's blood is tested.

Protein: A linear, unbranched polymer composed of twenty different chemical units called amino acids (Table 2.1). Proteins are the most diverse and functionally important macromolecules in our bodies.

Provider: See Healthcare provider.

Recessive (or autosomal recessive) single gene disorder: A single gene disorder in which both copies of a gene, located on one of the autosomes, are dysfunctional in affected individuals. With recessive disorders, both parents are healthy but have one functional and one dysfunctional gene. Each child of such a couple has an independent, 25% chance of inheriting both dysfunctional genes and being affected (Figure 4.4C). Contrast with dominant and X-linked single gene disorders.

Recombination (or meiotic recombination; also called crossing over): The exchange of DNA between pairs of chromosome DNA molecules of one type during meiosis (Figure 4.1). Recombination ensures that the chromosomes inherited by children are mosaics of the grandparental chromosomes (Figure 4.2).

Reference genome sequence: See Human reference genome sequence.

Regulation of genes: The body's control of gene function. Many genes are active (used to produce the coded protein) only in specific tissues. Genes may also be active only at particular times during our lives, and can be used to produce widely varying *amounts* of protein.

Replication: The copying of a DNA molecule to produce two nearly identical daughter molecules.

Reproductive planning: The process of planning pregnancies such that aneuploidies, single gene disorders and other genetic abnormalities are avoided in the child.

Ribosomes: Tiny molecular machines located in the cytoplasm that bind mRNA and manufacture proteins. Ribosomes are a complex of about 80 different proteins and 4 RNA molecules. The RNA molecules in ribosomes are different from mRNA.

RNA: Ribonucleic acid. A linear, unbranched polymer composed of four chemical units called nucleotides. The structure of RNA is very similar to DNA, except that RNA is single stranded, uses the nucleotide uridine instead of thymidine, and uses the sugar ribose instead of deoxyribose.

Risk index: A score determined by combining individual DNA risk factors *plus* environmental factors to predict the likelihood that a person will develop a specific genetically complex disorder or trait. Contrast with polygenic index.

Sequence variant: See Variant.

Sequencing: The process of determining the order of nucleotides along a DNA molecule. Powerful, inexpensive methods have been developed to sequence DNA.

Sex chromosomes: The X and Y chromosomes. Compare with autosomes.

Single-gene disorders (also called monogenic or Mendelian (after Gregor Mendel) disorders): A disorder or disability primarily caused by one or two variants in a single gene. Thousands of single-gene disorders are known.

Somatic cells: Any cell within our bodies other than the germ cells. Examples include blood, brain, skin and muscle cells. Mutations within somatic cells are not passed on to children.

Sperm cell: The male germ cell produced in the testes that is capable of fertilizing an egg cell.

Splicing: The process of removal of introns from an initial RNA transcript to produce a mature messenger RNA (mRNA) molecule.

Spontaneous abortion: Termination of a pregnancy that occurs naturally without any type of intervention. Spontaneous abortions can occur at any stage of pregnancy and are common, occurring in roughly 60% of all pregnancies. Spontaneous abortions that occur up to about the 20th week of pregnancy are called miscarriages. Those that occur after about the 20th week are called stillbirths. Contrast with elective abortion.

Stillbirth: A spontaneous abortion that occurs during the second half of normal pregnancy (more than about 20 weeks).

Structural mutation/variant (also called copy number mutation/variant): A mutation or variant involving the deletion or insertion of more than 50 nucleotides of DNA.

Symptom: See Clinical feature.

Synonymous mutation/variant: A mutation or variant within an exon that does *not* alter the amino acid sequence of the coded protein. Most synonymous mutations/variants are neutral or close to neutral.

Tandem repeats: Nucleotide sequences that are repeated one after another along a segment of a DNA molecule. The repeated unit can range in length anywhere from a single nucleotide to many thousands of nucleotides. Examples are shown in Figure 3.1.

Transcription: The process of producing RNA molecules from DNA templates. Transcription takes place in the cell nucleus.

Translation: The process of producing protein molecules from mRNA templates. Translation takes place in the cell cytoplasm.

Triploidy: Three copies of *every* chromosome in an individual instead of the normal two copies.

Trisomy: Three copies of a *single* chromosome in an individual instead of the normal two copies. An example is trisomy 21, also called Down syndrome (Figure 3.6).

Variable expressivity: When individuals (such as a pair of siblings) have exactly the same deleterious variant(s) for a single gene disorder yet have different disease courses and/or severities. Such individuals also often have different sets of clinical features.

Variant (also called sequence variant or DNA sequence variant): A difference between an individual's DNA sequence and the human reference sequence. In most cases, variant refers to sequences with less than 50% population frequency. Variants arise from mutations. Variants can be deleterious, neutral or advantageous.

X chromosome: One of the sex chromosomes. Females have two X chromosomes, and males one (Figures 2.13, 2.14, and 2.17).

X-linked single gene disorder: A single gene disorder in which the dysfunctional gene is located on the X chromosome. For the great majority of X-linked disorders, affected males greatly outnumber affected females. Typically, carrier females with one functional and one dysfunctional gene are unaffected or only mildly affected. However, because they have only one X chromosome, males who inherit the dysfunctional gene from their mothers will be fully affected. Each son of a carrier mother has an independent, 50% chance of inheriting the dysfunctional gene and being affected. Contrast with dominant and recessive single gene disorders.

Y chromosome: One of the sex chromosomes. Males have one Y chromosome, and females none (Figures 2.13, 2.14, and 2.17). Y chromosomes are inherited solely from fathers to sons.

REFERENCES

REFERENCES

Part I Genetics and Healthcare

Chapter 1 Introduction

1-1. Vogel F, Motulsky AG. Human Genetics, Problems and Approaches. Springer. Third Edition. 1996.

1-2. Mange E, Mange AP. Basic Human Genetics. Second Edition. Sinauer Associates. 1999.

1-3. Nussbaum RL, McInnes RR, Willard HF. Thompson and Thompson Genetics in Medicine. Eighth Edition. Elsevier. 2016.

1-4. Read A, Donnai D. New Clinical Genetics. Fourth Edition. Scion Publishers. 2020.

1-5. Penrose LS. Outline of Human Genetics. Third Edition. Heinemann Educational Books. 1973.

1-6. Osborne J, Pimentel D. Science, misinformation, and the role of education. Science. 378:246-248, 2022. PubMed ID: 36264815. [Open]

1-7. Weber JL. Suggestions for Long Term Goals for Humanity. Ten Generations Press. 2026. (Chapter 6).

Chapter 2 Human Genetics Basics

2-1. Amaral P et al. The status of the human gene catalogue. Nature. 622(7981):41-47, 2023. PubMed ID: 37794265. [Open]

2-2. Uhlén M et al. Tissue-based map of the human proteome. Science. 347(6220):1260419, 2015. PubMed ID: 25613900. [Open]

2-3. Sjöstedt E et al. An atlas of the protein-coding genes in the human, pig, and mouse brain. Science. 367(6482):eaay5947, 2020. PubMed ID: 32139519. [Open]

2-4. Chen W, Smeekens JM, Wu R. Systematic study of the dynamics and half-lives of newly synthesized proteins in human cells. Chemical Science. 7(2):1393-1400, 2016. PubMed ID: 29910897. (especially Figure 4) [Open]

2-5. Crick F. Central dogma of molecular biology. Nature. 227(5258):561-563, 1970. PubMed ID: 4913914.

2-6. Hatton IA et al. The human cell count and size distribution. Proceedings National Academy of Sciences USA. 120(39):e2303077120, 2023. PubMed ID: 37722043. [Open]

2-7. Zamenhof S, Brawerman G, Chargaff E. On the desoxypentose nucleic acids from several microorganisms. Biochimica et Biophysica Acta. 9(4):402-405, 1952. PubMed ID: 12997511.

2-8. Watson JD, Crick FH. Molecular structure of nucleic acids; a structure for deoxyribose nucleic acid. Nature. 171(4356):737-738, 1953. PubMed ID: 13054692.

2-9. Makani J et al. From Mendel to a Mendelian disorder: towards a cure for sickle cell disease. Nature Reviews Genetics. 23(7):389-390, 2022. PubMed ID: 35595849.

2-10. Raudenska M et al. Johann Gregor Mendel: the victory of statistics over human imagination. European Journal of Human Genetics. 31(7):744-748, 2023. PubMed ID: 36755104. [Open]

Chapter 3 Mutation

3-1. Hall LL, Smith KP, Lawrence JB. Emerging Functions of the Repeat Genome in Nuclear Structure: A View from the Human Karyotype. Annual Review of Genomics and Human Genetics. 26(1):45-75, 2025. PubMed ID: 40440481. [Open]

3-2. Sun JX et al. A direct characterization of human mutation based on microsatellites. Nature Genetics. 44(10):1161-1165, 2012. PubMed ID: 22922873. [Open]

3-3. Marcus R. What Is Huntington Disease? Journal of the American Medical Association. 330(10):1014, 2023. PubMed ID: 37603337. [Open]

3-4. Caron NS, Wright GEB, Hayden MR. Huntington Disease. 1998 Oct 23 [updated 2020 Jun 11]. In: Adam MP, Feldman J, Mirzaa GM, Pagon

RA, Wallace SE, Bean LJH, Gripp KW, Amemiya A, editors. GeneReviews® [Internet]. Seattle (WA): University of Washington, Seattle; 1993–2023. PubMed ID: 20301482. [Open]

3-5. Starlard-Davenport A. Complexity within simplicity: Exploring the multifactorial nature of sickle cell disease. American Journal of Human Genetics. 112(7):1499-1503, 2025. PubMed ID: 40482641. [Open]

3-6. Struewing JP et al. The carrier frequency of the *BRCA1* 185delAG mutation is approximately 1 percent in Ashkenazi Jewish individuals. Nature Genetics. 11(2):198-200, 1995. PubMed ID: 7550349.

3-7. Weijerman ME, de Winter JP. Clinical practice. The care of children with Down syndrome. European Journal of Pediatrics. 169(12):1445-1452, 2010. PubMed ID: 20632187. [Open]

3-8. Milholland B et al. Differences between germline and somatic mutation rates in humans and mice. Nature Communications. 8:15183, 2017. PubMed ID: 28485371. [Open]

3-9. Moore L et al. The mutational landscape of human somatic and germline cells. Nature. 597(7876):381-386, 2021. PubMed ID: 34433962.

3-10. Qian F, Germino GG. “Mistakes happen”: somatic mutation and disease. American Journal of Human Genetics. 61(5):1000-1005, 1997. PubMed ID: 9345111. [Open]

3-11. Akers AL et al. Biallelic somatic and germline mutations in cerebral cavernous malformations (CCMs): evidence for a two-hit mechanism of CCM pathogenesis. Human Molecular Genetics. 18(5):919-930, 2009. PubMed ID: 19088123. [Open]

3-12. Youssoufian H, Pyeritz RE. Mechanisms and consequences of somatic mosaicism in humans. Nature Reviews Genetics. 3(10):748-758, 2002. PubMed ID: 12360233.

3-13. Solís-Moruno M et al. Somatic genetic variation in healthy tissue and non-cancer diseases. European Journal of Human Genetics. 31(1):48-54, 2023. PubMed ID: 36289407. [Open]

3-14. Miller MB, Reed HC, Walsh CA. Brain Somatic Mutation in Aging and Alzheimer’s Disease. Annual Reviews of Genomics and Human Genetics. 22:239-256, 2021. PubMed ID: 33979534. [Open]

3-15. Chung C et al. Comprehensive multi-omic profiling of somatic mutations in malformations of cortical development. Nature Genetics. 55(2):209-220, 2023. PubMed ID: 36635388. [Open]

3-16. Maury EA et al. Schizophrenia-associated somatic copy-number variants from 12,834 cases reveal recurrent *NRXN1* and *ABCB11* disruptions. Cell Genomics. 3(8):100356, 2023. PubMed ID: 37601975. [Open]

3-17. Waldvogel SM, Posey JE, Goodell MA. Human embryonic genetic mosaicism and its effects on development and disease. Nature Reviews Genetics. 25(10):698-714, 2024. PubMed ID: 38605218. [Open]

3-18. Weber JL, Wong C. Mutation of human short tandem repeats. Human Molecular Genetics. 2(8):1123-1128, 1993. PubMed ID: 8401493.

3-19. Mitra I et al. Patterns of de novo tandem repeat mutations and their role in autism. Nature. 589(7841):246-250, 2021. PubMed ID: 33442040. [Open]

3-20. Jónsson H et al. Parental influence on human germline de novo mutations in 1,548 trios from Iceland. Nature. 549(7673):519-522, 2017. PubMed ID: 28959963.

3-21. Turner TN et al. Genomic Patterns of De Novo Mutation in Simplex Autism. Cell. 171(3):710-722, 2017. PubMed ID: 28965761. [Open]

3-22. Goldmann JM, Veltman JA, Gilissen C. De Novo Mutations Reflect Development and Aging of the Human Germline. Trends in Genetics. 35(11):828-839, 2019. PubMed ID: 31610893. [Open]

3-23. Noyes MD et al. Familial long-read sequencing increases yield of de novo mutations. American Journal of Human Genetics. 109(4):631-646, 2022. PubMed ID: 35290762. [Open]

3-24. Belyeu JR et al. De novo structural mutation rates and gamete-of-origin biases revealed through genome sequencing of 2,396 families. American Journal of Human Genetics. 108(4):597-607, 2021. PubMed ID: 33675682. [Open]

3-25. Porubsky D et al. Human de novo mutation rates from a four-generation pedigree reference. Nature. 643(8071):427-436, 2025. PubMed ID: 40269156. [Open]

3-26. Nussbaum RL, McInnes RR, Willard HF. Thompson and Thompson Genetics in Medicine. Eighth Edition. Elsevier. 2016. (Chapters 5 and 6)

3-27. Nagaoka SI, Hassold TJ, Hunt PA. Human aneuploidy: mechanisms and new insights into an age-old problem. Nature Reviews Genetics. 13(7):493-504, 2012. PubMed ID: 22705668. [Open]

3-28. Halgren C et al. Risks and Recommendations in Prenatally Detected De Novo Balanced Chromosomal Rearrangements from Assessment of Long-Term Outcomes. American Journal of Human Genetics. 102(6):1090-1103, 2018. PubMed ID: 29805044. [Open]

3-29. Francioli LC et al. Genome-wide patterns and properties of de novo mutations in humans. Nature Genetics. 47(7):822-826, 2015. PubMed ID: 25985141. [Open]

3-30. Rodriguez-Galindo M et al. Germline de novo mutation rates on exons versus introns in humans. Nature Communications. 11(1):3304, 2020. PubMed ID: 32620809. [Open]

3-31. Qiao L et al. Common variants increase risk for congenital diaphragmatic hernia within the context of de novo variants. American Journal of Human Genetics. 111(11):2362-2381, 2024. PubMed ID: 39332409. (Figure S3) [Open]

3-32. Kryukov GV, Pennacchio LA, Sunyaev SR. Most rare missense alleles are deleterious in humans: implications for complex disease and association studies. American Journal of Human Genetics. 80(4):727-739, 2007. PubMed ID: 17357078. [Open]

3-33. Eyre-Walker A, Keightley PD. The distribution of fitness effects of new mutations. Nature Reviews Genetics. 8(8):610-618, 2007. PubMed ID: 17637733.

3-34. Dobzhansky T. Genetics and the Origin of Species. Encyclopedia Britannica Inc. 1937.

3-35. Stenson PD et al. The Human Gene Mutation Database (HGMD®): optimizing its use in a clinical diagnostic or research setting. Human Genetics. 139(10):1197-1207, 2020. PubMed ID: 32596782. [Open]

3-36. Vallender EJ, Lahn BT. Positive selection on the human genome. Human Molecular Genetics. 13 Spec No 2:R245-254, 2004. PubMed ID: 15358731.

3-37. Findlay GM. Linking genome variants to disease: scalable approaches to test the functional impact of human mutations. Human Molecular Genetics. 30(R2):R187-R197, 2021. PubMed ID: 34338757. [Open]

3-38. Fowler DM et al. An Atlas of Variant Effects to understand the genome at nucleotide resolution. Genome Biology. 24(1):147, 2023. PubMed ID: 37394429. [Open]

3-39. Findlay GM et al. Accurate classification of *BRCA1* variants with saturation genome editing. Nature. 562(7726):217-222, 2018. PubMed ID: 30209399.

3-40. Radford EJ et al. Saturation genome editing of *DDX3X* clarifies pathogenicity of germline and somatic variation. Nature Communications. 14(1):7702, 2023. PubMed ID: 38057330. [Open]

3-41. Deciphering Developmental Disorders Study. Prevalence and architecture of de novo mutations in developmental disorders. Nature. 542(7642):433-438, 2017. PubMed ID: 28135719. [Open]

3-42. Kaplanis J et al. Evidence for 28 genetic disorders discovered by combining healthcare and research data. Nature. 586(7831):757-762, 2020. PubMed ID: 33057194. [Open]

3-43. Iossifov I et al. The contribution of de novo coding mutations to autism spectrum disorder. Nature. 515(7526):216-221, 2014. PubMed ID: 25363768. [Open]

3-44. Turner TN et al. Genomic Patterns of De Novo Mutation in Simplex Autism. Cell. 171(3):710-722, 2017. PubMed ID: 28965761. [Open]

3-45. Wang T, Zhao PA, Eichler EE. Rare variants and the oligogenic architecture of autism. Trends in Genetics. 38(9):895-903, 2022. PubMed ID: 35410794. [Open]

3-46. Trost B et al. Genomic architecture of autism from comprehensive whole-genome sequence annotation. Cell. 185(23):4409-4427, 2022. PubMed ID: 36368308. [Open]

3-47. King DG. Evolving a favorable distribution for mutation effects. Trends in Genetics. 40(10):819-821, 2024. PubMed ID: 39278786.

3-48. Hawkes G et al. Whole-genome sequencing analysis identifies rare, large-effect noncoding variants and regulatory regions associated with circulating protein levels. Nature Genetics. 57(3):626-634, 2025. PubMed ID: 39994471. [Open]

3-49. Kosicki M et al. In vivo mapping of mutagenesis sensitivity of human enhancers. Nature. 643(8072):839-846, 2025. PubMed ID: 40533554. [Open]

3-50. Ségurel L, Bon C. On the Evolution of Lactase Persistence in Humans. Annual Review of Genomics and Human Genetics. 18:297-319, 2017. PubMed ID: 28426286. [Open]

3-51. Crow JF. Basic Concepts in Population, Quantitative, and Evolutionary Genetics. W. H. Freeman and Company. 1986.

3-52. Hartl DL, Clark AG. Principles of Population Genetics. Third Edition. Sinauer Associates. 1997.

3-53. Kimura M. The neutral theory of molecular evolution and the world view of the neutralists. Genome. 31(1):24-31, 1989. PubMed ID: 2687096.

3-54. Darwin C. On the Origin of Species. First Edition. Penguin Classics. 1859.

3-55. Kingsley DM. From atoms to traits. Scientific American. 300(1):52-59, 2009. PubMed ID: 19186749. [Open]

3-56. Patterson N et al. Large-scale migration into Britain during the Middle to Late Bronze Age. Nature. 601(7894):588-594, 2022. PubMed ID: 34937049. [Open]

3-57. Mangelsdorf PC, Macneish RS, Galinat WC. Domestication of Corn. Science. 143(3606):538-545, 1964. PubMed ID: 17815644. [Open]

3-58. Doebley JF, Gaut BS, Smith BD. The molecular genetics of crop domestication. Cell. 127(7):1309-1321, 2006. PubMed ID: 17190597. [Open]

3-59. Yang N et al. Two teosintes made modern maize. Science. 382(6674):eadg8940, 2023. PubMed ID: 38033071. [Open]

3-60. Perri AR et al. Dog domestication and the dual dispersal of people and dogs into the Americas. Proceedings National Academy of Sciences USA. 118(6):e2010083118, 2021. PubMed ID: 33495362. [Open]

Chapter 4 Inheritance of Chromosomes, Variants, and Traits

4-1. Nussbaum RL, McInnes RR, Willard HF. Thompson and Thompson Genetics in Medicine. Eighth Edition. Elsevier. 2016.

4-2. Relethford JH. Human Population Genetics. Wiley-Blackwell. 2012.

4-3. Crow JF. Basic Concepts in Population, Quantitative, and Evolutionary Genetics. W. H. Freeman and Company. 1986. (Chapter 5).

4-4. Visscher PM, Hill WG, Wray NR. Heritability in the genomics era--concepts and misconceptions. Nature Reviews Genetics. 9(4):255-266, 2008. PubMed ID: 18319743.

4-5. Polderman TJ et al. Meta-analysis of the heritability of human traits based on fifty years of twin studies. Nature Genetics. 47(7):702-709, 2015. PubMed ID: 25985137.

4-6. Wang K et al. Classification of common human diseases derived from shared genetic and environmental determinants. Nature Genetics. 49(9):1319-1325, 2017. PubMed ID: 28783162. [Open]

4-7. Plomin R. Blueprint, How DNA Makes Us Wo We Are. MIT Press. 2018.

4-8. van Dongen J et al. The continuing value of twin studies in the omics era. Nature Reviews Genetics. 13(9):640-653, 2012. PubMed ID: 22847273.

4-9. Bouchard TJ Jr et al. Sources of human psychological differences: the Minnesota Study of Twins Reared Apart. Science. 250(4978):223-228, 1990. PubMed ID: 2218526. [Open]

4-10. Bicknell LS, Hirschhorn JN, Savarirayan R. The genetic basis of human height. Nature Reviews Genetics. 26(9):604-619, 2025. PubMed ID: 40189669.

4-11. Bouchard C. Genetics of Obesity: What We Have Learned Over Decades of Research. Obesity (Silver Spring). 29(5):802-820, 2021. PubMed ID: 33899337. [Open]

4-12. Pang H et al. The missing heritability in type 1 diabetes. Diabetes, Obesity and Metabolism. 24(10):1901-1911, 2022. PubMed ID: 35603907. [Open]

4-13. Ali O. Genetics of type 2 diabetes. World Journal of Diabetes. 4(4):114-123, 2013. PubMed ID: 23961321. [Open]

4-14. Devlin B, Daniels M, Roeder K. The heritability of IQ. Nature. 388(6641):468-471, 1997. PubMed ID: 9242404.

4-15. McClearn GE et al. Substantial genetic influence on cognitive abilities in twins 80 or more years old. Science. 276(5318):1560-1563, 1997. PubMed ID: 9171059. [Open]

4-16. Deary IJ, Spinath FM, Bates TC. Genetics of intelligence. European Journal of Human Genetics. 14(6):690-700, 2006. PubMed ID: 16721405.

4-17. Mucci LA et al. Familial Risk and Heritability of Cancer Among Twins in Nordic Countries. Journal of the American Medical Association. 315(1):68-76, 2016. PubMed ID: 26746459. [Open]

4-18. Sampson JN et al. Analysis of Heritability and Shared Heritability Based on Genome-Wide Association Studies for Thirteen Cancer Types. Journal of the National Cancer Institute. 107(12):djv279, 2015. PubMed ID: 26464424. [Open]

4-19. Nurhayu W et al. Handedness heritability in industrialized and nonindustrialized societies. Heredity (Edinb). 124(2):313-324, 2020. PubMed ID: 31602008. [Open]

4-20. Ocklenburg S et al. Genetics of human handedness: microtubules and beyond. Trends in Genetics. 41(6):497-505, 2025. PubMed ID: 39894687. [Open]

4-21. Boone PM et al. X-linked competition - implications for human development and disease. Nature Reviews Genetics. 26(8):571-580, 2025. PubMed ID: 40355603.

4-22. Muller HJ. Progress and prospects in human genetics. American Journal of Human Genetics. 1(1):1-18, 1949. PubMed ID: 17948379. [Open]

4-23. Fairbrother U et al. Genetics of Severe Obesity. Current Diabetes Reports. 18(10):85, 2018. PubMed ID: 30121879. [Open]

4-24. Bamshad MJ, Nickerson DA, Chong JX. Mendelian Gene Discovery: Fast and Furious with No End in Sight. American Journal of Human Genetics. 105(3):448-455, 2019. PubMed ID: 31491408. [Open]

4-25. Haendel M et al. How many rare diseases are there? Nature Reviews Drug Discovery. 19(2):77-78, 2020. PubMed ID: 32020066. [Open]

4-26. Azaiez H et al. Genomic Landscape and Mutational Signatures of Deafness-Associated Genes. American Journal of Human Genetics. 103(4):484-497, 2018. PubMed ID: 30245029. [Open]

4-27. Antoniou A et al. Average risks of breast and ovarian cancer associated with *BRCA1* or *BRCA2* mutations detected in case Series unselected for family history: a combined analysis of 22 studies. American Journal of Human Genetics. 72(5):1117-1130, 2003. PubMed ID: 12677558. [Open]

4-28. Li S et al. Cancer Risks Associated with *BRCA1* and *BRCA2* Pathogenic Variants. Journal of Clinical Oncology. 40(14):1529-1541, 2022. PubMed ID: 35077220. [Open]

4-29. Qian J et al. *APOE*-related risk of mild cognitive impairment and dementia for prevention trials: An analysis of four cohorts. PLoS Medicine. 14(3):e1002254, 2017. PubMed ID: 28323826. [Open]

4-30. Bird TD. Alzheimer Disease Overview. 1998 Oct 23 [updated 2018 Dec 20]. In: Adam MP, Feldman J, Mirzaa GM, Pagon RA, Wallace SE, Bean LJH, Gripp KW, Amemiya A, editors. GeneReviews® [Internet]. Seattle (WA): University of Washington, Seattle; 1993–2024. PubMed ID: 20301340. [Open]

4-31. Wellcome Trust Case Control Consortium. Genome-wide association study of 14,000 cases of seven common diseases and 3,000 shared controls. Nature. 447(7145):661-678, 2007. PubMed ID: 17554300. [Open]

4-32. Abdellaoui A et al. 15 years of GWAS discovery: Realizing the promise. American Journal of Human Genetics. 110(2):179-194, 2023. PubMed ID: 36634672. [Open]

4-33. Scriver CR. The *PAH* gene, phenylketonuria, and a paradigm shift. Human Mutation. 28(9):831-845, 2007. PubMed ID: 17443661. [Open]

4-34. Breen G. Common genetic variants contribute more to rare diseases than previously thought. Nature. 636(8042):304-305, 2024. PubMed ID: 39567806.

4-35. Huang QQ et al. Examining the role of common variants in rare neurodevelopmental conditions. Nature. 636(8042):404-411, 2024. PubMed ID: 39567701. [Open]

4-36. Fahed AC et al. Polygenic background modifies penetrance of monogenic variants for tier 1 genomic conditions. Nature Communications. 11(1):3635, 2020. PubMed ID: 32820175. [Open]

4-37. Riordan JD, Nadeau JH. From Peas to Disease: Modifier Genes, Network Resilience, and the Genetics of Health. American Journal of Human Genetics. 101(2):177-191, 2017. PubMed ID: 28777930. [Open]

4-38. Wang T, Zhao PA, Eichler EE. Rare variants and the oligogenic architecture of autism. Trends in Genetics. 38(9):895-903, 2022. PubMed ID: 35410794. [Open]

4-39. Falconer DS. The inheritance of liability to diseases with variable age of onset, with particular reference to diabetes mellitus. Annals of Human Genetics. 31(1):1-20, 1967. PubMed ID: 6056557.

Chapter 5 Human Evolutionary History and Population Genetics

5-1. Fortney R. Life: A Natural History of the First Four Billion Years of Life on Earth. Vintage Books. 1997.

5-2. Gee HA. A (Very) Short History of Life on Earth. St. Martin's Press. 2021.

5-3. Wu S et al. Genomes, fossils, and the concurrent rise of modern birds and flowering plants in the Late Cretaceous. Proceedings of the National Academy of Science USA. 121(8):e2319696121, 2024. PubMed ID: 38346181. [Open]

5-4. Scally A et al. Insights into hominid evolution from the gorilla genome sequence. Nature. 483(7388):169-175, 2012. PubMed ID: 22398555. [Open]

5-5. Rogers J, Gibbs RA. Comparative primate genomics: emerging patterns of genome content and dynamics. Nature Reviews Genetics. 15(5):347-359, 2014. PubMed ID: 24709753. [Open]

5-6. Roberts A, editor. Evolution, the Human Story. Penguin Random House. Second Edition. 2018.

5-7. Wood B, Harrison T. The evolutionary context of the first hominins. Nature. 470(7334):347-352, 2011. PubMed ID: 21331035.

5-8. Alemseged Z. Reappraising the palaeobiology of Australopithecus. Nature. 617(7959):45-54, 2023. PubMed ID: 37138108.

5-9. Lieberman DE. Palaeoanthropology: Homo floresiensis from head to toe. Nature. 459(7243):41-42, 2009. PubMed ID: 19424142.

5-10. Peyrégne S, Slon V, Kelso J. More than a decade of genetic research on the Denisovans. Nature Reviews Genetics. 25(2):83-103, 2024. PubMed ID: 37723347.

5-11. Higham T. The World Before Us. Yale University Press. 2021.

5-12. Nielsen R et al. Tracing the peopling of the world through genomics. Nature. 541(7637):302-310, 2017. PubMed ID: 28102248. [Open]

5-13. Skoglund P, Mathieson I. Ancient Genomics of Modern Humans: The First Decade. Annual Review of Genomics and Human Genetics. 19:381-404, 2018. PubMed ID: 29709204. [Open]

5-14. Bergström A et al. Origins of modern human ancestry. Nature. 590(7845):229-237, 2021. PubMed ID: 33568824.

5-15. Reich D. Who We Are and How We Got Here, Ancient DNA and the New Science of the Human Past. Pantheon Books. 2018.

5-16. Ragsdale AP et al. A weakly structured stem for human origins in Africa. Nature. 617(7962):755-763, 2023. PubMed ID: 37198480. [Open]

5-17. Pääbo S. Molecular cloning of Ancient Egyptian mummy DNA. Nature. 314(6012):644-645, 1985. PubMed ID: 3990798.

5-18. Dalén L et al. Deep-time paleogenomics and the limits of DNA survival. Science. 382(6666):48-53, 2023. PubMed ID: 37797036. [Open]

5-19. Ávila-Arcos MC, Raghavan M, Schlebusch C. Going local with ancient DNA: A review of human histories from regional perspectives. Science. 382(6666):53-58, 2023. PubMed ID: 37797024. [Open]

5-20. Meyer M et al. Nuclear DNA sequences from the Middle Pleistocene Sima de los Huesos hominins. Nature. 531(7595):504-507, 2016. PubMed ID: 26976447.

5-21. Green RE et al. A draft sequence of the Neandertal genome. Science. 328(5979):710-722, 2010. PubMed ID: 20448178. [Open]

5-22. Sankararaman S et al. The genomic landscape of Neanderthal ancestry in present-day humans. Nature. 507(7492):354-357, 2014. PubMed ID: 24476815. [Open]

5-23. Skov L et al. The nature of Neanderthal introgression revealed by 27,566 Icelandic genomes. Nature. 582(7810):78-83, 2020. PubMed ID: 32494067.

5-24. Iasi LNM et al. Neanderthal ancestry through time: Insights from genomes of ancient and present-day humans. Science. 386(6727):eadq3010, 2024. PubMed ID: 39666853. [Open]

5.25. Ongaro L, Huerta-Sanchez E. A history of multiple Denisovan introgression events in modern humans. Nature Genetics. 56(12):2612-2622, 2024. PubMed ID: 39501127.

5-26. Jorde LB, Watkins WS, Bamshad MJ. Population genomics: a bridge from evolutionary history to genetic medicine. Human Molecular Genetics. 10(20):2199-2207, 2001. PubMed ID: 11673402. [Open]

5-27. Kruglyak L, Nickerson DA. Variation is the spice of life. Nature Genetics. 27(3):234-236, 2001. PubMed ID: 11242096.

5-28. Morton NE, Crow JF, Muller HJ. An estimate of the mutational damage in man from data on consanguineous marriages. Proceedings of the National Academy of Sciences USA. 42(11):855-863, 1956. PubMed ID: 16589958. [Open]

5-29. Chong JX et al. A population-based study of autosomal-recessive disease-causing mutations in a founder population. American Journal of Human Genetics. 91(4):608-620, 2012. PubMed ID: 22981120. [Open]

5-30. Kirk EP et al. Nationwide, Couple-Based Genetic Carrier Screening. New England Journal of Medicine. 391(20):1877-1889, 2024. PubMed ID: 39565987. [Open]

5-31. Relethford JH. Human Population Genetics. Wiley-Blackwell. 2012.

5-32. Crow JF. Basic Concepts in Population, Quantitative, and Evolutionary Genetics. W. H. Freeman and Company. 1986. (Chapter 5).

5-33. Hartl DL, Clark AG. Principles of Population Genetics, Third Edition. Sinauer Associates. 1997.

5-34. Henn BM et al. Estimating the mutation load in human genomes. Nature Reviews Genetics. 16(6):333-343, 2015. PubMed ID: 25963372. [Open]

5-35. Pemberton TJ et al. Genomic patterns of homozygosity in worldwide human populations. American Journal of Human Genetics. 91(2):275-292, 2012. PubMed ID: 22883143. [Open]

5-36. Byrska-Bishop M et al. High-coverage whole-genome sequencing of the expanded 1000 Genomes Project cohort including 602 trios. Cell. 185(18):3426-3440, 2022. PubMed ID: 36055201. [Open]

5-37. Rosenberg NA et al. Genetic structure of human populations. Science. 298(5602):2381-2385, 2002. PubMed ID: 12493913. [Open]

Chapter 6 Applications of Clinical DNA Testing

6-1. Filchakova O et al. Review of COVID-19 testing and diagnostic methods. Talanta. 244:123409, 2022. PubMed ID: 35390680. [Open]

6-2. Hinrichs A et al. The ongoing evolution of UShER during the SARS-CoV-2 pandemic. Nature Genetics. 56(1):4-7, 2024. PubMed ID: 38155331.

6-3. Telenti A, Hodcroft EB, Robertson DL. The Evolution and Biology of SARS-CoV-2 Variants. Cold Spring Harbor Perspectives in Medicine. 12(5):a041390, 2022. PubMed ID: 35444005. [Open]

6-4. Fitzgerald RC et al. The future of early cancer detection. Nature Medicine. 28(4):666-677, 2022. PubMed ID: 35440720.

6-5. Crosby D et al. Early detection of cancer. Science. 375(6586):eaay9040, 2022. PubMed ID: 35298272. [Open]

6-6. Micalizzi DS, Sequist LV, Haber DA. Deploying blood-based cancer screening. Science. 383(6681):368-370, 2024. PubMed ID: 38271495. [Open]

6-7. Friedman AA et al. Precision medicine for cancer with next-generation functional diagnostics. Nature Reviews Cancer. 15(12):747-756, 2015. PubMed ID: 26536825. [Open]

6-8. Dupont CA et al. Druggable genome and precision medicine in cancer: current challenges. FEBS Journal. 288(21):6142-6158, 2021. PubMed ID: 33626231. [Open]

6-9. Middleton G et al. A state-of-the-art review of stratified medicine in cancer: towards a future precision medicine strategy in cancer. Annals of Oncology. 33(2):143-157, 2022. PubMed ID: 34808340. [Open]

6-10. Denny JC, Collins FS. Precision medicine in 2030-seven ways to transform healthcare. Cell. 184(6):1415-1419, 2021. PubMed ID: 33740447. [Open]

6-11. National Academy of Medicine, Consensus Study Report: Improving Diagnosis in Healthcare (https://nap.nationalacademies.org/catalog/21794/improving-diagnosis-in-health-care). 2015.

6-12. Newman-Toker DE et al. Burden of serious harms from diagnostic error in the USA. BMJ Quality and Safety. 33(2):109-120, 2024. PubMed ID: 37460118. [Open]

6-13. Marini C, Giardino M. Novel treatments in epilepsy guided by genetic diagnosis. British Journal of Clinical Pharmacology. 88(6):2539-2551, 2022. PubMed ID: 34778987. [Open]

6-14. Zimmern V, Minassian B, Korff C. A Review of Targeted Therapies for Monogenic Epilepsy Syndromes. Frontiers in Neurology. 13:829116, 2022. PubMed ID: 35250833. [Open]

6-15. Gospe SM Jr. Pyridoxine-Dependent Epilepsy – *ALDH7A1*. 2001 [updated 2022]. In: Adam MP, Feldman J, Mirzaa GM, Pagon RA, Wallace SE, Bean LJH, Gripp KW, Amemiya A, editors. GeneReviews® [Internet]. Seattle (WA): University of Washington, Seattle; 1993–2024. PubMed ID: 20301659. [Open]

6-16. Kurahashi H, Hirose S. Autosomal Dominant Sleep-Related Hypermotor (Hyperkinetic) Epilepsy. 2002 [updated 2023]. In: Adam

MP, Feldman J, Mirzaa GM, Pagon RA, Wallace SE, Bean LJH, Gripp KW, Amemiya A, editors. GeneReviews® [Internet]. Seattle (WA): University of Washington, Seattle; 1993–2024. PubMed ID: 20301348. [Open]

6-17. Miceli F et al. *KCNQ2*-Related Disorders. 2010 [updated 2022]. In: Adam MP, Feldman J, Mirzaa GM, Pagon RA, Wallace SE, Bean LJH, Gripp KW, Amemiya A, editors. GeneReviews® [Internet]. Seattle (WA): University of Washington, Seattle; 1993–2024. PubMed ID: 20437616. [Open]

6-18. Gertler T et al. *KCNT1*-Related Epilepsy. 2018. In: Adam MP, Feldman J, Mirzaa GM, Pagon RA, Wallace SE, Bean LJH, Gripp KW, Amemiya A, editors. GeneReviews® [Internet]. Seattle (WA): University of Washington, Seattle; 1993–2024. PubMed ID: 30234941. [Open]

6-19. Yang K, Quiroz V, Ebrahimi-Fakhari D. *PRRT2*-Related Disorder. 2018 [updated 2024]. In: Adam MP, Feldman J, Mirzaa GM, Pagon RA, Wallace SE, Amemiya A, editors. GeneReviews® [Internet]. Seattle (WA): University of Washington, Seattle; 1993–2025. PubMed ID: 29334453. [Open]

6-20. Miller IO, Sotero de Menezes MA. *SCN1A* Seizure Disorders. 2007 [updated 2022]. In: Adam MP, Feldman J, Mirzaa GM, Pagon RA, Wallace SE, Bean LJH, Gripp KW, Amemiya A, editors. GeneReviews® [Internet]. Seattle (WA): University of Washington, Seattle; 1993–2024. PubMed ID: 20301494. [Open]

6-21. Northrup H et al. Tuberous Sclerosis Complex. 1999 [updated 2024]. In: Adam MP, Feldman J, Mirzaa GM, Pagon RA, Wallace SE, Amemiya A, editors. GeneReviews® [Internet]. Seattle (WA): University of Washington, Seattle; 1993–2025. PubMed ID: 20301399. [Open]

6-22. Savant A et al. Cystic Fibrosis. 2001 [updated 2024]. In: Adam MP, Feldman J, Mirzaa GM, Pagon RA, Wallace SE, Amemiya A, editors. GeneReviews® [Internet]. Seattle (WA): University of Washington, Seattle; 1993–2025. PubMed ID: 20301428. [Open]

6-23. Jia S, Taylor-Cousar JL. Cystic Fibrosis Modulator Therapies. Annual Review of Medicine. 74:413-426, 2023. PubMed ID: 35973718. [Open]

6-24. Dunbar CE et al. Gene therapy comes of age. Science. 359(6372):eaan4672, 2018. PubMed ID: 29326244. [Open]

6-25. Martinez Velazquez LA, Ballios BG. The Next Generation of Molecular and Cellular Therapeutics for Inherited Retinal Disease. International Journal of Molecular Sciences. 22(21):11542, 2021. PubMed ID: 34768969. [Open]

6-26. Hoytema van Konijnenburg EMM et al. Treatable inherited metabolic disorders causing intellectual disability: 2021 review and digital app. Orphanet Journal of Rare Diseases. 16(1):170, 2021. PubMed ID: 33845862. [Open]

6-27. Savarese M et al. The genetic basis of undiagnosed muscular dystrophies and myopathies: Results from 504 patients. Neurology. 87(1):71-76, 2016. PubMed ID: 27281536. [Open]

6-28. Alankarage D et al. Identification of clinically actionable variants from genome sequencing of families with congenital heart disease. Genetics in Medicine. 21(5):1111-1120, 2019. PubMed ID: 30293987. [Open]

6-29. Shearer AE et al. Genetic Hearing Loss Overview. 1999 [updated 2025]. In: Adam MP, Feldman J, Mirzaa GM, Pagon RA, Wallace SE, Amemiya A, editors. GeneReviews® [Internet]. Seattle (WA): University of Washington, Seattle; 1993–2025. PubMed ID: 20301607. [Open]

6-30. Cocchi E, Nestor JG, Gharavi AG. Clinical Genetic Screening in Adult Patients with Kidney Disease. Clinical Journal of the American Society of Nephrology. 15(10):1497-1510, 2020. PubMed ID: 32646915. [Open]

6-31. Zhou X et al. Integrating de novo and inherited variants in 42,607 autism cases identifies mutations in new moderate-risk genes. Nature Genetics. 54(9):1305-1319, 2022. PubMed ID: 35982159. [Open]

6-32. Pandey R et al. A meta-analysis of diagnostic yield and clinical utility of genome and exome sequencing in pediatric rare and undiagnosed genetic diseases. Genetics in Medicine. 27(6):101398, 2025. PubMed ID: 40022598. [Open]

6-33. Kingsmore SF, Cole FS. The Role of Genome Sequencing in Neonatal Intensive Care Units. Annual Review of Genomics and Human Genetics. 23:427-448, 2022. PubMed ID: 35676073. [Open]

6-34. Goldin MR et al. Benefits and barriers to broad implementation of genomic sequencing in the NICU. American Journal of Human Genetics. 112(6):1270-1285, 2025. PubMed ID: 40367948. [Open]

6-35. Sardh E, Barbaro M. Acute Intermittent Porphyria. 2005 [updated 2024]. In: Adam MP, Feldman J, Mirzaa GM, Pagon RA, Wallace SE, Amemiya A, editors. GeneReviews® [Internet]. Seattle (WA): University of Washington, Seattle; 1993–2025. PubMed ID: 20301372. [Open]

6-36. Wenger BM et al. A genotype-first approach to exploring Mendelian cardiovascular traits with clear external manifestations. Genetics in Medicine. 23(1):94-102, 2021. PubMed ID: 32989268. [Open]

6-37. Klemenzdottir EO et al. A population-based survey of *FBN1* variants in Iceland reveals underdiagnosis of Marfan syndrome. European Journal of Human Genetics. 32(1):44-51, 2024. PubMed ID: 37684520. [Open]

6-38. Savatt JM et al. Genomic Screening at a Single Health System. JAMA Network Open. 8(3):e250917, 2025. PubMed ID: 40094662. [Open]

6-39. Brown TR et al. Genetic Testing Among Medicaid-Insured Children with Autism and Intellectual Disability. JAMA Network Open. 8(9):e2533518, 2025. PubMed ID: 40971157. [Open]

6-40. Tinker RJ et al. Diagnostic delay in monogenic disease: A scoping review. Genetics in Medicine. 26(4):101074, 2024. PubMed ID: 38243783. [Open]

6-41. Saeidian AH et al. Secondary ACMG and non-ACMG genetic findings in a multiethnic cohort of 16,713 pediatric participants. Genetics in Medicine. 26(11):101225, 2024. PubMed ID: 39096151. [Open]

6-42. Arnold G, Vockley J. Phenylalanine Hydroxylase Deficiency. 2000 [updated 2025]. In: Adam MP, Feldman J, Mirzaa GM, Pagon RA, Wallace SE, Amemiya A, editors. GeneReviews® [Internet]. Seattle (WA): University of Washington, Seattle; 1993–2025. PubMed ID: 20301677. [Open]

6-43. Idos G, Valle L. Lynch Syndrome. 2004 [updated 2021]. In: Adam MP, Feldman J, Mirzaa GM, Pagon RA, Wallace SE, Bean LJH, Gripp KW, Amemiya A, editors. GeneReviews® [Internet]. Seattle (WA): University of Washington, Seattle; 1993–2024. PubMed ID: 20301390. [Open]

6-44. Torkamani A, Wineinger NE, Topol EJ. The personal and clinical utility of polygenic risk scores. Nature Reviews Genetics. 19(9):581-590, 2018. PubMed ID: 29789686.

6-45. Cross B, Turner R, Pirmohamed M. Polygenic risk scores: An overview from bench to bedside for personalised medicine. Frontiers in Genetics. 13:1000667, 2022. PubMed ID: 36437929. [Open]

6-46. Abdellaoui A et al. 15 years of GWAS discovery: Realizing the promise. American Journal of Human Genetics. 110(2):179-194, 2023. PubMed ID: 36634672. [Open]

6-47. Yang X et al. Polygenic scores in cancer. Nature Reviews Cancer. 23(9):619-630, 2023. PubMed ID: 37479830.

6-48. Chang CH et al. Polygenic height prediction for the Han Chinese in Taiwan. NPJ Genomic Medicine. 10(1):7, 2025. PubMed ID: 39910149. [Open]

6-49. Detrois KE et al. Cross-biobank generalizability and accuracy of electronic health record-based predictors compared to polygenic scores. Nature Genetics. 57(9):2136-2145, 2025. PubMed ID: 40866628. [Open]

6-50. Mars N et al. Genome-wide risk prediction of common diseases across ancestries in one million people. Cell Genomics. 2(4), 2022. PubMed ID: 35591975. [Open]

6-51. McHugh JK et al. Assessment of a Polygenic Risk Score in Screening for Prostate Cancer. New England Journal of Medicine. 392(14):1406-1417, 2025. PubMed ID: 40214032. [Open]

6-52. Kullo IJ et al. The PRIMED Consortium: Reducing disparities in polygenic risk assessment. American Journal of Human Genetics. 111(12):2594-2606, 2024. PubMed ID: 39561770. [Open]

6-53. Vassy JL et al. The GenoVA study: Equitable implementation of a pragmatic randomized trial of polygenic-risk scoring in primary care. American Journal of Human Genetics. 110(11):1841-1852, 2023. PubMed ID: 37922883. [Open]

6-54. Lennon NJ et al. Selection, optimization and validation of ten chronic disease polygenic risk scores for clinical implementation in diverse US populations. Nature Medicine. 30(2):480-487, 2024. PubMed ID: 38374346. [Open]

6-55. Rao ND et al. Association of *HFE* genotypes with hemochromatosis-related phenotypes in the All of Us research program. Genetics in Medicine Open. 3:101959, 2025. PubMed ID: 40034140. [Open]

6-56. Dayan CM et al. Preventing type 1 diabetes in childhood. Science. 373(6554):506-510, 2021. PubMed ID: 34326231. [Open]

6-57. Kolovos A et al. Polygenic Risk Scores Driving Clinical Change in Glaucoma. Annual Review of Genomics and Human Genetics. 25(1):287-308, 2024. PubMed ID: 38599222. [Open]

6-58. Di Carlo E, Augustin AJ. Prevention of the Onset of Age-Related Macular Degeneration. Journal of Clinical Medicine. 10(15):3297, 2021. PubMed ID: 34362080. [Open]

6-59. Lombardo M, Serrao S, Lombardo G. Challenges in Age-Related Macular Degeneration: From Risk Factors to Novel Diagnostics and Prevention Strategies. Frontiers in Medicine (Lausanne). 9:887104, 2022. PubMed ID: 35733877. [Open]

6-60. Grimes KR et al. Effects of systemic drugs on the development and progression of age-related macular degeneration. Survey of Ophthalmology. 68(3):332-346, 2023. PubMed ID: 36731638.

6-61. Li Y et al. Cell-type-directed network-correcting combination therapy for Alzheimer's disease. Cell. 188(20):5516-5534, 2025. PubMed ID: 40695276. [Open]

6-62. Courade JP et al. The evolving landscape of Alzheimer's disease therapy: From Aβ to tau. Cell. 188(26):7337-7354, 2025. PubMed ID: 41448139. [Open]

6-63. Schalkamp AK et al. Wearable movement-tracking data identify Parkinson's disease years before clinical diagnosis. Nature Medicine. 29(8):2048-2056, 2023. PubMed ID: 37400639. [Open]

6-64. Shen Y Dominantly Inherited Alzheimer Network; Ibanez L, Sung YJ, Cruchaga C. CSF proteomics identifies early changes in autosomal dominant Alzheimer's disease. Cell. 187(22):6309-6326.e15, 2024. PubMed ID: 39332414. [Open]

6-65. Toro C, Shirvan L, Tifft C. *HEXA* Disorders. 2020. In: Adam MP, Feldman J, Mirzaa GM, Pagon RA, Wallace SE, Bean LJH, Gripp KW, Amemiya A, editors. GeneReviews® [Internet]. Seattle (WA): University of Washington, Seattle; 1993–2024. PubMed ID: 20301397. [Open]

6-66. Delatycki MB et al. International perspectives on the implementation of reproductive carrier screening. Prenatal Diagnosis. 40(3):301-310, 2020. PubMed ID: 31774570.

6-67. Ravitsky V et al. The Emergence and Global Spread of Noninvasive Prenatal Testing. Annual Review of Genomics and Human Genetics. 22:309-338, 2021. PubMed ID: 33848430. [Open]

6-68. de Graaf G et al. Estimation of the number of people with Down syndrome living in Canada. Genetics in Medicine. 27(7):101422, 2025. PubMed ID: 40196939.

6-69. Auwerx C et al. From pharmacogenetics to pharmaco-omics: Milestones and future directions. Human Genetics and Genomics Advances. 3(2):100100, 2022. PubMed ID: 35373152. [Open]

6-70. Haidar CE et al. Advancing Pharmacogenomics from Single Gene to Preemptive Testing. Annual Review of Genomics and Human Genetics. 23:449-473, 2022. PubMed ID: 35537468. [Open]

6-71. Pirmohamed M. Pharmacogenomics: current status and future perspectives. Nature Reviews Genetics. 24(6):350-362, 2023. PubMed ID: 36707729.

6-72. Schork NJ. Personalized medicine: Time for one-person trials. Nature. 520(7549):609-611, 2015. PubMed ID: 25925459.

6-73. Sultana J, Cutroneo P, Trifirò G. Clinical and economic burden of adverse drug reactions. Journal of Pharmacology and Pharmacotherapeutics. 4(Suppl 1):S73-77, 2013. PubMed ID: 24347988. [Open]

6-74. Insani WN et al. Prevalence of adverse drug reactions in the primary care setting: A systematic review and meta-analysis. PLoS One. 16(5):e0252161, 2021. PubMed ID: 34038474. [Open]

6-75. Doherty AS et al. Adverse drug reactions and associated patient characteristics in older community-dwelling adults: a 6-year prospective cohort study. British Journal of General Practice. 73(728):e211-e219, 2023. PubMed ID: 36823047. [Open]

6-76. McDermott JH et al. Rapid Point-of-Care Genotyping to Avoid Aminoglycoside-Induced Ototoxicity in Neonatal Intensive Care. JAMA Pediatrics. 176(5):486-492, 2022. PubMed ID: 35311942. [Open]

6-77. Oslin DW et al. Effect of Pharmacogenomic Testing for Drug-Gene Interactions on Medication Selection and Remission of Symptoms in Major Depressive Disorder: The PRIME Care Randomized Clinical Trial. Journal of the American Medical Association. 328(2):151-161, 2022. PubMed ID: 35819423. [Open]

6-78. Kang Z et al. Multigenetic Pharmacogenomics-Guided Treatment vs Treatment as Usual Among Hospitalized Men with Schizophrenia: A

Randomized Clinical Trial. JAMA Network Open. 6(10):e2335518, 2023. PubMed ID: 37801319. [Open]

6-79. Eadon MT et al. The INGENIOUS trial: Impact of pharmacogenetic testing on adverse events in a pragmatic clinical trial. The Pharmacogenomics Journal. 23(6):169-177, 2023. PubMed ID: 37689822. [Open]

6-80. Swen JJ et al. A 12-gene pharmacogenetic panel to prevent adverse drug reactions: an open-label, multicentre, controlled, cluster-randomised crossover implementation study. Lancet. 401(10374):347-356, 2023. PubMed ID: 36739136.

6-81. Abul-Husn NS, Kenny EE. Personalized Medicine and the Power of Electronic Health Records. Cell. 177(1):58-69, 2019. PubMed ID: 30901549. [Open]

6-82. Szustakowski JD et al. Advancing human genetics research and drug discovery through exome sequencing of the UK Biobank. Nature Genetics. 53(7):942-948, 2021. PubMed ID: 34183854.

6-83. Butler JM. The future of forensic DNA analysis. Philosophical Transactions of the Royal Society B. 370(1674):20140252, 2015. PMID: 26101278. [Open]

6-84. Biesecker LG et al. DNA identifications after the 9/11 World Trade Center attack. Science. 310(5751):1122-1123, 2005. PMID: 16293742. [Open]

6-85. Kishore A, Petrek M. Next-Generation Sequencing Based HLA Typing: Deciphering Immunogenetic Aspects of Sarcoidosis. Frontiers in Genetics. 9:503, 2018. PMID: 30410504. [Open]

6-86. Li HY, Guo K. Blood Group Testing. Frontiers in Medicine (Lausanne). 9:827619, 2022. PMID: 35223922. [Open]

Chapter 7 Clinical DNA Testing Factors

7-1. Alankarage D et al. Identification of clinically actionable variants from genome sequencing of families with congenital heart disease. Genetics in Medicine. 21(5):1111-1120, 2019. PubMed ID: 30293987. [Open]

7-2. Shearer AE et al. Genetic Hearing Loss Overview. 1999 [updated 2025]. In: Adam MP, Feldman J, Mirzaa GM, Pagon RA, Wallace SE, Amemiya A, editors. GeneReviews® [Internet]. Seattle (WA): University of Washington, Seattle; 1993–2025. PubMed ID: 20301607. [Open]

7-3. Cocchi E, Nestor JG, Gharavi AG. Clinical Genetic Screening in Adult Patients with Kidney Disease. Clinical Journal of the American Society of Nephrology. 15(10):1497-1510, 2020. PubMed ID: 32646915. [Open]

7-4. Zhou X et al. Integrating de novo and inherited variants in 42,607 autism cases identifies mutations in new moderate-risk genes. Nature Genetics. 54(9):1305-1319, 2022. PubMed ID: 35982159. [Open]

7-5. van der Sanden BPGH et al. The performance of genome sequencing as a first-tier test for neurodevelopmental disorders. European Journal of Human Genetics. 31(1):81-88, 2023. PubMed ID: 36114283. [Open]

7-6. Pandey R et al. A meta-analysis of diagnostic yield and clinical utility of genome and exome sequencing in pediatric rare and undiagnosed genetic diseases. Genetics in Medicine. 27(6):101398, 2025. PubMed ID: 40022598. [Open]

7-7. Jenkins BD et al. The 2019 US medical genetics workforce: a focus on clinical genetics. Genetics in Medicine. 23(8):1458-1464, 2021. PubMed ID: 33941882. [Open]

7-8. American Board of Medical Genetics and Genomics: Resources (https://www.abmgg.org/resources/sub-item-2/). 2025.

7-9. U.S. Bureau of Labor Statistics (https://www.bls.gov/oes/current/oes291215.htm). 2024.

7-10. National Society of Genetic Counselors (https://www.nsgc.org/POLICY/Genetic-Counselor-Workforce). 2021.

7-11. Schaaf CP. Genetic counseling and the role of genetic counselors in the United States. Medizinische Genetik. 33(1):29-34, 2021. PubMed ID: 38836204. [Open]

7-12. Ormond KE et al. The global status of genetic counselors in 2023: What has changed in the past 5 years? Genetics in Medicine Open. 2(Suppl 2):101887, 2024. PubMed ID: 39712962. [Open]

7-13. Maiese DR et al. The 2019 medical genetics workforce: A focus on laboratory geneticists. Genetics in Medicine. 25(6):100834, 2023. PubMed ID: 36999554. [Open]

7-14. Abacan M et al. The Global State of the Genetic Counseling Profession. European Journal of Human Genetics. 27(2):183-197, 2019. PubMed ID: 30291341. [Open]

7-15. McEwen JE et al. The Ethical, Legal, and Social Implications Program of the National Human Genome Research Institute: reflections on an ongoing experiment. Annual Review of Genomics and Human Genetics. 15:481-505, 2014. PubMed ID: 24773317.

7-16. Broadstock M, Michie S, Marteau T. Psychological consequences of predictive genetic testing: a systematic review. European Journal of Human Genetics. 8(10):731-738, 2000. PubMed ID: 11039571. [Open]

7-17. Heshka JT et al. A systematic review of perceived risks, psychological and behavioral impacts of genetic testing. Genetics in Medicine. 10(1):19-32, 2008. PubMed ID: 18197053. [Open]

7-18. Hamilton JG, Lobel M, Moyer A. Emotional distress following genetic testing for hereditary breast and ovarian cancer: a meta-analytic review. Health Psychology. 28(4):510-518, 2009. PubMed ID: 19594276. [Open]

7-19. Joly Y et al. Looking Beyond GINA: Policy Approaches to Address Genetic Discrimination. Annual Review of Genomics and Human Genetics. 21:491-507, 2020. PubMed ID: 31961723. [Open]

7-20. Collins FS, Morgan M, Patrinos A. The Human Genome Project: lessons from large-scale biology. Science. 300(5617):286-290, 2003. PubMed ID: 12690187. [Open]

7-21. McDonald-McGinn DM et al. 22q11.2 Deletion Syndrome. 1999 [updated 2025]. In: Adam MP, Feldman J, Mirzaa GM, Pagon RA, Wallace SE, Amemiya A, editors. GeneReviews® [Internet]. Seattle (WA): University of Washington, Seattle; 1993–2025. PubMed ID: 20301696. [Open]

7-22. Weiss KH, Schilsky M. Wilson Disease. 1999 [updated 2023]. In: Adam MP, Feldman J, Mirzaa GM, Pagon RA, Wallace SE, Bean LJH, Gripp KW, Amemiya A, editors. GeneReviews® [Internet]. Seattle (WA): University of Washington, Seattle; 1993–2024. PubMed ID: 20301685. [Open]

7-23. Yang Y et al. Molecular findings among patients referred for clinical whole-exome sequencing. Journal of the American Medical Association. 312(18):1870-1879, 2014. PubMed ID: 25326635. [Open]

7-24. Farwell KD et al. Enhanced utility of family-centered diagnostic exome sequencing with inheritance model-based analysis: results from 500 unselected families with undiagnosed genetic conditions. Genetics in Medicine. 17(7):578-586, 2015. PubMed ID: 25356970. [Open]

7-25. Posey JE et al. Molecular diagnostic experience of whole-exome sequencing in adult patients. Genetics in Medicine. 18(7):678-685, 2016. PubMed ID: 26633545. [Open]

7-26. Lavelle TA et al. Cost-effectiveness of exome and genome sequencing for children with rare and undiagnosed conditions. Genetics in Medicine. 24(6):1349-1361, 2022. PubMed ID: 35396982. [Open]

7-27. Davidson AL et al. The clinical utility and costs of whole-genome sequencing to detect cancer susceptibility variants-a multi-site prospective cohort study. Genome Medicine. 15(1):74, 2023. PubMed ID: 37723522. [Open]

7-28. Glaubitz R et al. The cost of the diagnostic odyssey of patients with suspected rare diseases. Orphanet Journal of Rare Diseases. 20(1):222, 2025. PubMed ID: 40349051. [Open]

7-29. Farnaes L et al. Rapid whole-genome sequencing decreases infant morbidity and cost of hospitalization. NPJ Genomic Medicine. 3:10, 2018. PubMed ID: 29644095. [Open]

7-30. van der Sanden BPGH et al. The performance of genome sequencing as a first-tier test for neurodevelopmental disorders. European Journal of Human Genetics. 31(1):81-88, 2023. PubMed ID: 36114283. [Open]

7-31. Chung CCY et al. Meta-analysis of the diagnostic and clinical utility of exome and genome sequencing in pediatric and adult patients with rare diseases across diverse populations. Genetics in Medicine. 25(9):100896, 2023. PubMed ID: 37191093. [Open]

7-32. Lowther C et al. Systematic evaluation of genome sequencing for the diagnostic assessment of autism spectrum disorder and fetal structural anomalies. American Journal of Human Genetics. 110(9):1454-1469, 2023. PubMed ID: 37595579. [Open]

7-33. Nurchis MC et al. Cost-Effectiveness of Whole-Genome vs Whole-Exome Sequencing Among Children with Suspected Genetic Disorders. JAMA Network Open. 7(1):e2353514, 2024. PubMed ID: 38277144. [Open]

7-34. Schobers G et al. Uncovering recessive alleles in rare Mendelian disorders by genome sequencing of 174 individuals with monoallelic pathogenic variants. European Journal of Human Genetics. 33(1):56-64, 2025. PubMed ID: 39333430.

7-35. American Medical Association: Trends in Health Care Spending. (https://www.ama-assn.org/about/research/trends-health-care-spending). 2025.

7-36. Li C et al. Cost-effectiveness of genome-wide sequencing for unexplained developmental disabilities and multiple congenital anomalies. Genetics in Medicine. 23(3):451-460, 2021. PubMed ID: 33110268. [Open]

7-37. Lavelle TA et al. Cost-effectiveness of exome and genome sequencing for children with rare and undiagnosed conditions. Genetics in Medicine. 24(6):1349-1361, 2022. PubMed ID: 35396982. [Open]

7-38. Kingsmore SF, Cole FS. The Role of Genome Sequencing in Neonatal Intensive Care Units. Annual Review of Genomics and Human Genetics. 23:427-448, 2022. PubMed ID: 35676073. [Open]

7-39. Wu Y et al. Genomic testing for suspected monogenic kidney disease in children and adults: A health economic evaluation. Genetics in Medicine. 25(11):100942, 2023. PubMed ID: 37489581. [Open]

7-40. Davidson AL et al. The clinical utility and costs of whole-genome sequencing to detect cancer susceptibility variants-a multi-site prospective cohort study. Genome Medicine. 15(1):74, 2023. PubMed ID: 37723522. [Open]

7-41. Guzauskas GF et al. Population Genomic Screening for Three Common Hereditary Conditions: A Cost-Effectiveness Analysis. Annals of Internal Medicine. 176(5):585-595, 2023. PubMed ID: 37155986.

7-42. Spencer SJ et al. Cost-effectiveness of population-wide genomic screening for familial hypercholesterolemia in the United States. Journal of Clinical Lipidology. 16(5):667-675, 2022. PubMed ID: 35961838. [Open]

7-43. Law CK et al. Long-term cost-effectiveness of a melanoma prevention program using genomic risk information compared with standard prevention advice in Australia. Genetics in Medicine. 25(12):100970, 2023. PubMed ID: 37658729. [Open]

7-44. Schofield D et al. Cost-effectiveness of population-based expanded reproductive carrier screening for genetic diseases in Australia: a microsimulation analysis. European Journal of Human Genetics. 33(9):1180-1187, 2025 PubMed ID: 40240435. [Open]

7-45. Carreiro-Lewandowski E. Newborn screening: an overview. Clinical Laboratory Science. 15(4):229-238, 2002. PubMed ID: 12776783.

7-46. Watson MS, Lloyd-Puryear MA, Howell RR. The Progress and Future of US Newborn Screening. International Journal of Neonatal Screening. 8(3):41, 2022. PubMed ID: 35892471. [Open]

7-47. Stark Z, Scott RH. Genomic newborn screening for rare diseases. Nature Reviews Genetics. 24(11):755-766, 2023. PubMed ID: 37386126.

7-48. Therrell BL et al. Current Status of Newborn Bloodspot Screening Worldwide 2024: A Comprehensive Review of Recent Activities (2020-2023). International Journal of Neonatal Screening. 10(2):38, 2024. PubMed ID: 38920845. [Open]

7-49. Arnold G, Vockley J. Phenylalanine Hydroxylase Deficiency. 2000 [updated 2025]. In: Adam MP, Feldman J, Mirzaa GM, Pagon RA, Wallace SE, Amemiya A, editors. GeneReviews® [Internet]. Seattle (WA): University of Washington, Seattle; 1993–2025. PubMed ID: 20301677. [Open]

7-50. Miles JH. Autism spectrum disorders--a genetics review. Genetics in Medicine. 13(4):278-294, 2011. PubMed ID: 21358411. [Open]

7-51. Lord C et al. Autism spectrum disorder. Lancet. 392(10146):508-520, 2018. PubMed ID: 30078460. [Open]

7-52. Zeidan J et al. Global prevalence of autism: A systematic review update. Autism Research. 15(5):778-790, 2022. PubMed ID: 35238171. [Open]

7-53. van 't Hof M et al. Age at autism spectrum disorder diagnosis: A systematic review and meta-analysis from 2012 to 2019. Autism. 25(4):862-873, 2021. PubMed ID: 33213190. [Open]

7-54. Perochon S et al. Early detection of autism using digital behavioral phenotyping. Nature Medicine. 29(10):2489-2497, 2023. PubMed ID: 37783967. [Open]

7-55. Franz L et al. Early intervention for very young children with or at high likelihood for autism spectrum disorder: An overview of reviews. Developmental Medicine and Child Neurology. 64(9):1063-1076, 2022. PubMed ID: 35582893. [Open]

7-56. Iossifov I et al. The contribution of de novo coding mutations to autism spectrum disorder. Nature. 515(7526):216-221, 2014. PubMed ID: 25363768. [Open]

7-57. Turner TN et al. Genomic Patterns of De Novo Mutation in Simplex Autism. Cell. 171(3):710-722, 2017. PubMed ID: 28965761. [Open]

7-58. Wang T, Zhao PA, Eichler EE. Rare variants and the oligogenic architecture of autism. Trends in Genetics. 38(9):895-903, 2022. PubMed ID: 35410794. [Open]

7-59. Wenger AM et al. Systematic reanalysis of clinical exome data yields additional diagnoses: implications for providers. Genetics in Medicine. 19(2):209-214, 2017. PubMed ID: 27441994. [Open]

7-60. Bartolomaeus T et al. Re-evaluation and re-analysis of 152 research exomes five years after the initial report reveals clinically relevant changes in 18. European Journal of Human Genetics. 31(10):1154-1164, 2023. PubMed ID: 37460657. [Open]

7-61. Appelbaum PS et al. Practical considerations for reinterpretation of individual genetic variants. Genetics in Medicine. 25(5):100801, 2023. PubMed ID: 36748709. [Open]

7-62. Gudmundsson S et al. Variant interpretation using population databases: Lessons from gnomAD. Human Mutation. 43(8):1012-1030, 2022. PubMed ID: 34859531. [Open]

7-63. Tan VY, Timpson NJ. The UK Biobank: A Shining Example of Genome-Wide Association Study Science with the Power to Detect the Murky Complications of Real-World Epidemiology. Annual Review of Genomics and Human Genetics. 23:569-589, 2022. PubMed ID: 35508184. [Open]

7-64. Mayo KR et al. The *All of Us* Data and Research Center: Creating a Secure, Scalable, and Sustainable Ecosystem for Biomedical Research. Annual Review of Biomedical Data Science. 6:443-464, 2023. PubMed ID: 37561600. [Open]

7-65. Gallagher CS, Ginsburg GS, Musick A. Biobanking with genetics shapes precision medicine and global health. Nature Reviews Genetics. 26(3):191-202, 2025. PubMed ID: 39567741.

7-66. Eisfeldt J, Ek M, Nordenskjöld M, Lindstrand A. Toward clinical long-read genome sequencing for rare diseases. Nature Genetics. 57(6):1334-1343, 2025. PubMed ID: 40335760.

Chapter 8 Summary Part 1

8-1. Hodson R. Precision medicine. Nature. 537(7619):S49, 2016. PubMed ID: 27602738.

8-2. Ginsburg GS, Phillips KA. Precision Medicine: From Science To Value. Health Affairs (Millwood). 37(5):694-701, 2018. PubMed ID: 29733705. [Open]

8-3. Denny JC, Collins FS. Precision medicine in 2030-seven ways to transform healthcare. Cell. 184(6):1415-1419, 2021. PubMed ID: 33740447. [Open]

8-4. Stark Z et al. Integrating Genomics into Healthcare: A Global Responsibility. American Journal of Human Genetics. 104(1):13-20, 2019. PubMed ID: 30609404. [Open]

8-5. Kovanda A, Zimani AN, Peterlin B. How to design a national genomic project-a systematic review of active projects. Human Genomics. 15(1):20, 2021. PubMed ID: 33761998. [Open]

8-6. Williams MS. Population Screening in Health Systems. Annual Reviews of Genomics and Human Genetics. 23:549-567, 2022. PubMed ID: 35175786. [Open]

8-7. Stark Z, Scott RH. Genomic newborn screening for rare diseases. Nature Reviews Genetics. 24(11):755-766, 2023. PubMed ID: 37386126. [Open]

8-08. Ziegler A, Chung WK. Genomic newborn screening: exploring opportunities and navigating pitfalls while ensuring inclusivity. Trends in Genetics. 41(8):631-634, 2025. PubMed ID: 40300982.

8-09. UK Biobank. (https://www.ukbiobank.ac.uk/). 2025.

8-10. UK NHS 10 Year Health Plan (https://www.gov.uk/government/publications/10-year-health-plan-for-england-fit-for-the-future). 2025.

8-11. Wong E et al. The Singapore National Precision Medicine Strategy. Nature Genetics. 55(2):178-186, 2023. PubMed ID: 36658435.

8-12. Precision Health Research, Singapore (PRECISE) (https://www.npm.sg/about-us/our-story/). 2025.

8-13. Kurki MI et al. FinnGen provides genetic insights from a well-phenotyped isolated population. Nature. 613(7944):508-518, 2023. PubMed ID: 36653562. [Open]

8-14. Palotie, A, Kaunisto M, Daly M. The FinnGen project, a unique resource for genetic discoveries. Finska Läkaresällskapets Handlingar (Proceedings of the Finnish Medical Association). 183(1):50-55, 2023. (https://fls.fi/wp-content/uploads/2023/05/10_123_Handlingar6.pdf).

Part II Management of the Gene Pool

Chapter 9 Introduction to Part II

9-1. Neel JV. Medicine's genetic horizons. Annals of Internal Medicine 49(2):472-476, 1958. PubMed ID: 13571834.

9-2. Weber JL. Suggestions for Long Term Goals for Humanity. Ten Generations Press. 2026.

Chapter 10 Prenatal Development and Prenatal Testing

10-1. Nussbaum RL, McInnes RR, Willard HF. Thompson and Thompson Genetics in Medicine. Eighth Edition. Elsevier. 2016. (Chapter 14)

10-2. Tang WW et al. Specification and epigenetic programming of the human germ line. Nature Reviews Genetics. 17(10):585-600, 2016. PubMed ID: 27573372.

10-3. Larose H et al. Gametogenesis: A journey from inception to conception. Current Topics in Developmental Biology. 132:257-310, 2019. PubMed ID: 30797511. [Open]

10-4. Moris N. Stem cells used to model a two-week-old human embryo. Nature. 622(7983):469-470, 2023. PubMed ID: 37848524.

10-5. Boklage CE. Survival probability of human conceptions from fertilization to term. International Journal of Fertility. 35(2):75, 79-80, 81-94, 1990. PubMed ID: 1970983.

10-6. Macklon NS, Geraedts JP, Fauser BC. Conception to ongoing pregnancy: the 'black box' of early pregnancy loss. Human Reproduction Update. 8(4):333-343, 2002. PubMed ID: 12206468.

10-7. Quenby S et al. Miscarriage matters: the epidemiological, physical, psychological, and economic costs of early pregnancy loss. Lancet. 397(10285):1658-1667, 2021. PubMed ID: 33915094.

10-8. Brosens JJ et al. Maternal selection of human embryos in early gestation: Insights from recurrent miscarriage. Seminars in Cell and Developmental Biology. 131:14-24, 2022. PubMed ID: 35094946. [Open]

10-9. Nagaoka SI, Hassold TJ, Hunt PA. Human aneuploidy: mechanisms and new insights into an age-old problem. Nature Reviews Genetics. 13(7):493-504, 2012. PubMed ID: 22705668. [Open]

10-10. Essers R et al. Prevalence of chromosomal alterations in first-trimester spontaneous pregnancy loss. Nature Medicine. 29(12):3233-3242, 2023. PubMed ID: 37996709. [Open]

10-11. Arnadottir GA et al. Sequence diversity lost in early pregnancy. Nature. 642(8068):672-681, 2025. PubMed ID: 40399685. [Open]

10-12. Nussbaum RL, McInnes RR, Willard HF. Thompson and Thompson Genetics in Medicine. Eighth Edition. Elsevier. 2016. (Chapter 5)

10-13. Krivega M, Stiefel CM, Storchova Z. Consequences of chromosome gain: A new view on trisomy syndromes. American Journal of Human Genetics. 109(12):2126-2140, 2022. PubMed ID: 36459979. [Open]

10-14. Ludorf KL et al. Long-Term Survival Among Children With Trisomy 13 and Trisomy 18 by Cytogenetic Status. JAMA Network Open. Open. 8(9):e2529885, 2025. PubMed ID: 40920381. [Open]

10-15. Hook EB, Warburton D. Turner syndrome revisited: review of new data supports the hypothesis that all viable 45,X cases are cryptic mosaics with a rescue cell line, implying an origin by mitotic loss. Human Genetics. 133(4):417-424, 2014. PubMed ID: 24477775.

10-16. Frederiksen LE et al. Maternal age and the risk of fetal aneuploidy: a nationwide cohort study of more than 500,000 singleton pregnancies in Denmark from 2008 to 2017. Acta Obstetricia et Gynecologica Scandinavica. 103(2):351-359, 2024. [Open]

10-17. StatPearls Publishing. Chorionic Villus Sampling. (https://www.ncbi.nlm.nih.gov/books/NBK563301/). 2025.

10-18. StatPearls Publishing. Amniocentesis. (https://www.ncbi.nlm.nih.gov/books/NBK559247/). 2023.

10-19. Cohen JL et al. Advancing precision care in pregnancy through a treatable fetal findings list. American Journal of Human Genetics. 112(6):1251-1269, 2025. PubMed ID: 40209713. [Open]

10-20. Guttmacher Institute, State Bans on Abortion Throughout Pregnancy. (https://www.guttmacher.org/state-policy/explore/state-policies-abortion-bans#:~:text=12%20states%20have%20a%20total,some%20point%20after%2018%20weeks). 2025.

10-21. Kalantari S et al. Termination of pregnancy for fetal malformations and severe genetic disorders: what are the laws in Europe. European Journal of Human Genetics. 33:151-153, 2025. PubMed ID: 39026088.

10-22. Lo YM et al. Presence of fetal DNA in maternal plasma and serum. Lancet. 350(9076):485-487, 1997. PubMed ID: 9274585.

10-23. Lo YMD. Discovery of Cell-Free Fetal DNA in Maternal Blood and Development of Noninvasive Prenatal Testing: 2022 Lasker-DeBakey Clinical Medical Research Award. Journal of the American Medical Association. 328(13):1293-1294, 2022. PubMed ID: 36170057. [Open]

10-24. Ravitsky V et al. The Emergence and Global Spread of Noninvasive Prenatal Testing. Annual Reviews of Genomics and Human Genetics. 22:309-338, 2021. PubMed ID: 33848430. [Open]

10-25. Moufarrej MN et al. Noninvasive Prenatal Testing Using Circulating DNA and RNA: Advances, Challenges, and Possibilities. Annual Review of Biomedical Data Science. 6:397-418, 2023. PubMed ID: 37196360. [Open]

10-26. Lun FM et al. Microfluidics digital PCR reveals a higher than expected fraction of fetal DNA in maternal plasma. Clinical Chemistry. 54(10):1664-1672, 2008. PubMed ID: 18703764. [Open]

10-27. Taylor-Phillips S et al. Accuracy of non-invasive prenatal testing using cell-free DNA for detection of Down, Edwards and Patau syndromes: a systematic review and meta-analysis. British Medical Journal Open. 6(1):e010002, 2016. PubMed ID: 26781507. [Open]

10-28. Dungan JS et al. Electronic address: documents@acmg.net. Noninvasive prenatal screening (NIPS) for fetal chromosome abnormalities in a general-risk population: An evidence-based clinical guideline of the American College of Medical Genetics and Genomics (ACMG). Genetics in Medicine. 25(2):100336, 2023. PubMed ID: 36524989. [Open]

10-29. Alberry M et al. Free fetal DNA in maternal plasma in anembryonic pregnancies: confirmation that the origin is the trophoblast. Prenatal Diagnosis. 27(5):415-418, 2007. PubMed ID: 17286310.

10-30. American College of Obstetricians and Gynecologists' Committee on Practice Bulletins—Obstetrics; Committee on Genetics; Society for Maternal-Fetal Medicine. Screening for Fetal Chromosomal Abnormalities: ACOG Practice Bulletin, Number 226. Obstetrics and Gynecology. 136(4):e48-e69, 2020. PubMed ID: 32804883.

10-31. Zhang J et al. Prospective prenatal cell-free DNA screening for genetic conditions of heterogenous etiologies. Nature Medicine. 30(2):470-479, 2024. PubMed ID: 38253798.

10-32. Zhong LPW, Chiu RWK. The Next Frontier in Noninvasive Prenatal Diagnostics: Cell-Free Fetal DNA Analysis for Monogenic Disease Assessment. Annual Reviews of Genomics and Human Genetics. 23:413-425, 2022. PubMed ID: 35316613. [Open]

10-33. Hoskovec J et al. Maternal carrier screening with single-gene NIPS provides accurate fetal risk assessments for recessive conditions. Genetics in Medicine. 25(2):100334, 2023. PubMed ID: 36454238. [Open]

10-34. Brand H et al. High-Resolution and Noninvasive Fetal Exome Screening. New England Journal of Medicine. 389(21):2014-2016, 2023. PubMed ID: 37991862. [Open]

10-35. Miceikaitė I et al. Comprehensive Noninvasive Fetal Screening by Deep Trio-Exome Sequencing. New England Journal of Medicine. 389(21):2017-2019, 2023. PubMed ID: 37991863. [Open]

10-36. Fan HC et al. Non-invasive prenatal measurement of the fetal genome. Nature. 487(7407):320-324, 2012. PubMed ID: 22763444. [Open]

10-37. Gadsbøll K et al. Current use of noninvasive prenatal testing in Europe, Australia and the USA: A graphical presentation. Acta Obstetricia et Gynecologica Scandinavica. 99(6):722-730, 2020. PubMed ID: 32176318. [Open]

10-38. Van Den Bogaert K et al. Outcome of publicly funded nationwide first-tier noninvasive prenatal screening. Genetics in Medicine. 23(6):1137-1142, 2021. PubMed ID: 33564150. [Open]

10-39. de Graaf G et al. Estimation of the number of people with Down syndrome living in Canada. Genetics in Medicine. 27(7):101422, 2025. PubMed ID: 40196939.

Chapter 11 Genetic Reproductive Planning

11-1. Online Mendelian Inheritance in Man. (https://www.omim.org/). 2025.

11-2. Ferreira CR. The burden of rare diseases. American Journal of Medical Genetics Part A. 179(6):885-892, 2019. PubMed ID: 30883013.

11-3. Nguengang Wakap S et al. Estimating cumulative point prevalence of rare diseases: analysis of the Orphanet database. European Journal of Human Genetics. 28(2):165-173, 2020. PubMed ID: 31527858. [Open]

11-4. Gunne E et al. An estimate of the cumulative paediatric prevalence of rare diseases in Ireland and comment on the literature. European Journal of Human Genetics. 30(11):1211-1215, 2022. PubMed ID: 35853949. [Open]

11-5. Haendel M et al. How many rare diseases are there? Nature Reviews Drug Discovery. 19(2):77-78, 2020. PubMed ID: 32020066. [Open]

11-6. Chung CCY et al. Socio-economic costs of rare diseases and the risk of financial hardship: a cross-sectional study. The Lancet Regional Health - West Pacific. 34:100711, 2023. PubMed ID: 37283971. [Open]

11-7. Costa T, Scriver CR, Childs B. The effect of Mendelian disease on human health: a measurement. American Journal of Medical Genetics. 21(2):231-242, 1985. PubMed ID: 4014310.

11-8. Owen MJ et al. Reclassification of the Etiology of Infant Mortality with Whole-Genome Sequencing. JAMA Network Open. 6(2):e2254069, 2023. PubMed ID: 36757698. [Open]

11-9. Walker CE et al. The collective impact of rare diseases in Western Australia: an estimate using a population-based cohort. Genetics in Medicine. 19(5):546-552, 2017. PubMed ID: 27657686. [Open]

11-10. Marshall DA et al. Direct health-care costs for children diagnosed with genetic diseases are significantly higher than for children with other chronic diseases. Genetics in Medicine. 21(5):1049-1057, 2019. PubMed ID: 30245512. [Open]

11-11. Gonzaludo N et al. Estimating the burden and economic impact of pediatric genetic disease. Genetics in Medicine. 21(8):1781-1789, 2019. PubMed ID: 30568310. [Open]

11-12. Navarrete-Opazo AA et al. Can you hear us now? The impact of health-care utilization by rare disease patients in the United States. Genetics in Medicine. 23(11):2194-2201, 2021. PubMed ID: 34183788. [Open]

11-13. Kish AM, Newcombe PA, Haslam DM. Working and caring for a child with chronic illness: A review of current literature. Child: Care, Health and Development. 44(3):343-354, 2018. PubMed ID: 29341191.

11-14. Wu Y et al. Parental health spillover effects of paediatric rare genetic conditions. Quality of Life Research. 29(9):2445-2454, 2020. PubMed ID: 32266555.

11-15. Boettcher J et al. Being the Pillar for Children with Rare Diseases-A Systematic Review on Parental Quality of Life. International Journal of

Environmental Research and Public Health. 18(9):4993, 2021. PubMed ID: 34066738. [Open]

11-16. Gallop K et al. Impact of developmental and epileptic encephalopathies on caregivers: A literature review. Epilepsy & Behavior. 124:108324, 2021. PubMed ID: 34607217. [Open]

11-17. Yang G et al. The national economic burden of rare disease in the United States in 2019. Orphanet Journal of Rare Disease. 17(1):163, 2022. PubMed ID: 35414039. [Open]

11-18. Grasemann H, Ratjen F. Cystic Fibrosis. New England Journal of Medicine. 389(18):1693-1707, 2023. PubMed ID: 37913507.

11-19. Guo J et al. Current prices versus minimum costs of production for CFTR modulators. Journal of Cystic Fibrosis. 21(5):866-872, 2022. PubMed ID: 35440408. [Open]

11-20. Cystic Fibrosis Foundation. (https://www.cff.org/about-us/importance-cost-and-affordability-people-cf). 2024.

11-21. Bulaklak K, Gersbach CA. The once and future gene therapy. Nature Communications. 11(1):5820, 2020. PubMed ID: 33199717. [Open]

11-22. Mendell JR et al. Current Clinical Applications of In Vivo Gene Therapy with AAVs. Molecular Therapy. 29(2):464-488, 2021. PubMed ID: 33309881. [Open]

11-23. Wong CH et al. The estimated annual financial impact of gene therapy in the United States. Gene Therapy. 30(10-11):761-773, 2023. PubMed ID: 37935855. [Open]

11-24. Scotti C, Aiuti A, Naldini L. Challenges and solutions to the sustainability of gene and cell therapies. Nature Reviews Genetics. 26(7):437-438, 2025. PubMed ID: 40065154.

11-25. Milunsky A. Heredity and Your Family's Health. Johns Hopkins University Press. 1989. (Preface).

11-26. Maranto G. Quest for Perfection, The Drive to Breed Better Human Beings. iUniverse.com Inc. 1996. (Chapter 9)

11-27. Ombelet W, Van Robays J. Artificial insemination history: hurdles and milestones. Facts Views and Vision in ObGyn. 7(2):137-143, 2015. PubMed ID: 26175891. [Open]

11-28. Arocho R, Lozano EB, Halpern CT. Estimates of donated sperm use in the United States: National Survey of Family Growth 1995-2017. Fertility and Sterility. 112(4):718-723, 2019. PubMed ID: 31371048. [Open]

1-29. Jain M, Singh M. Assisted Reproductive Technology (ART) Techniques. In: StatPearls [Internet]. Treasure Island (FL): StatPearls Publishing, 2023. PubMed ID: 35015434. [Open]

11-30. Lafuente-Funes S et al. Egg donation in the age of vitrification: A study of egg providers' perceptions and experiences in the UK, Belgium and Spain. Sociology of Health and Illness. 45(2):259-278, 2023. PubMed ID: 36444136. [Open]

11-31. Boulet SL et al. Trends in use of and reproductive outcomes associated with intracytoplasmic sperm injection. Journal of the American Medical Association. 313(3):255-263, 2015. PubMed ID: 25602996. [Open]

11-31. Haddad M et al. Thoughts on the popularity of ICSI. Journal of Assisted Reproduction and Genetics. 38(1):101-123, 2021. PubMed ID: 33155089. [Open]

11-32. Berntsen S et al. IVF versus ICSI in patients without severe male factor infertility: a randomized clinical trial. Nature Medicine. 31(6):1939-1948, 2025. PubMed ID: 40217077. [Open]

11-33. Shandley LM, DeSantis CE, Lee JC, Kawwass JF, Hipp HS. Trends and Outcomes of Assisted Reproductive Technology Cycles Using a Gestational Carrier Between 2014 and 2020. Journal of the American Medical Association. 330(17):1691-1694, 2023. PubMed ID: 37851614. [Open]

11-34. Herbert M, Choudhary M, Zander-Fox D. Assisted reproductive technologies at the nexus of fertility treatment and disease prevention. Science. 380(6641):164-167, 2023. PubMed ID: 37053308. [Open]

11-35. U.S. Centers for Disease Control and Prevention. ART Success Rates. (https://www.cdc.gov/art/success-rates/?CDC_AAref_Val=https://www.cdc.gov/art/artdata/index.html). 2025.

11-36. Society for Assisted Reproductive Technology. National Summary Report (https://sartcorsonline.com/CSR/PublicSnapshotReport?ClinicPKID=0&reportingYear=2022). 2022.

11-37. Brison DR. IVF children and healthy aging. Nature Medicine. 28(12):2476-2477, 2022. PubMed ID: 36536255.

11-38. Luke B et al. The risks of birth defects and childhood cancer with conception by assisted reproductive technology. Human Reproduction. 37(11):2672-2689, 2022. PubMed ID: 36112004. [Open]

11-39. Stern JE et al. Assisted reproductive technology treatment increases obstetric and neonatal risks over that of the underlying infertility diagnosis. Fertility and Sterility. 117(6):1223-1234, 2022. PubMed ID: 35397876. [Open]

11-40. Toro C, Shirvan L, Tifft C. *HEXA* Disorders. 2020. In: Adam MP, Feldman J, Mirzaa GM, Pagon RA, Wallace SE, Bean LJH, Gripp KW, Amemiya A, editors. GeneReviews® [Internet]. Seattle (WA): University of Washington, Seattle; 1993–2024. PubMed ID: 20301397. [Open]

11-41. Dor Yeshorim. (https://doryeshorim.org/). 2025.

11-42. Anderson S. Current State of Dor Yeshorim, Expanded Carrier, and Newborn Screening: Benefits and Limitations. American Journal of Maternal Child Nursing. 48(5):266-272, 2023. PubMed ID: 37326600.

11-43. Fridman H et al. The landscape of autosomal-recessive pathogenic variants in European populations reveals phenotype-specific effects. American Journal of Human Genetics. 108(4):608-619, 2021. PubMed ID: 33740458. [Open]

11-44. Schmitz MJ et al. Leveraging diverse genomic data to guide equitable carrier screening: Insights from gnomAD v.4.1.0. American Journal of Human Genetics. 112(1):181-195, 2025. PubMed ID: 39615480. [Open]

11-45. Hotakainen R et al. Estimation of carrier frequencies of autosomal and X-linked recessive genetic conditions based on gnomAD v4.0 data in different ancestries. Genetics in Medicine. 27(1):101304, 2025. PubMed ID: 39492094. [Open]

11-46. Meyer MN et al. Public views on polygenic screening of embryos. Science. 379(6632):541-543, 2023. PubMed ID: 36758092. [Open]

11-47. Turley P et al. Problems with Using Polygenic Scores to Select Embryos. New England Journal of Medicine. 385(1):78-86, 2021. PubMed ID: 34192436. [Open]

11-48. Grebe TA et al. Clinical utility of polygenic risk scores for embryo selection: A points to consider statement of the American College of Medical Genetics and Genomics (ACMG). Genetics in Medicine. 26(4):101052, 2024. PubMed ID: 38393332. [Open]

11-49. Furrer RA et al. Public Attitudes, Interests, and Concerns Regarding Polygenic Embryo Screening. JAMA Network Open. 7(5):e2410832, 2024. PubMed ID: 38743425. [Open]

11-50. Bearak J et al. Unintended pregnancy and abortion by income, region, and the legal status of abortion: estimates from a comprehensive model for 1990-2019. Lancet Global Health. 8(9):e1152-e1161, 2020. PubMed ID: 32710833. [Open]

11-51. Bearak JM et al. Country-specific estimates of unintended pregnancy and abortion incidence: a global comparative analysis of levels in 2015-2019. BMJ Global Health. 7(3):e007151, 2022. PubMed ID: 35332057. [Open]

11-52. Delatycki MB et al. International perspectives on the implementation of reproductive carrier screening. Prenatal Diagnosis. 40(3):301-310, 2020. PubMed ID: 31774570.

11-53. Singer A, Sagi-Dain L. Impact of a national genetic carrier-screening program for reproductive purposes. Acta Obstetricia et Gynecologica Scandinavica. 99(6):802-808, 2020. PubMed ID: 32242916. [Open]

11-54. Ravitsky V et al. The Emergence and Global Spread of Noninvasive Prenatal Testing. Annual Review of Genomics and Human Genetics. 22:309-338, 2021. PubMed ID: 33848430. [Open]

11-55. de Graaf G et al. Estimation of the number of people with Down syndrome living in Canada. Genetics in Medicine. 27(7):101422, 2025. PubMed ID: 40196939.

11-56. Lin G et al. Genetic rare disease prevention and control: Family-based screening and reproductive interventions in Changsha. HGG Advances. 6(4):100496, 2025. PubMed ID: 40831058. [Open]

11-57. Nov-Klaiman T, Horn R, Raz A. More of the same? Israel's expanded carrier screening for cystic fibrosis. European Journal of Human Genetics. 33(12):1555-1557, 2025. PubMed ID: 40281080. [Open]

11-58. Relethford JH. Human Population Genetics. Wiley-Blackwell. 2012. (Figure 6.4)

11-59. Crow JF. Basic Concepts in Population, Quantitative, and Evolutionary Genetics. W. H. Freeman and Company. 1986. (Figures 4.1 and 4.2)

Chapter 12 Genetics of Intelligence

12-1. Gould SJ. The Mismeasure of Man. W.W. Norton and Co. 1981; revised 1996.

12-2. Flynn JR. Are We Getting Smarter? Rising IQ in the Twenty-First Century. Cambridge University Press. 2012.

12-3. Pinker S. The Blank Slate, The Modern Denial of Human Nature. Viking, Penguin Group. 2002.

12-4. Mackintosh NJ. IQ and Human Intelligence. Oxford University Press. Second Edition, 2010.

12-5. Herrnstein RJ, Murray C. The Bell Curve. Simon and Schuster. 1994.

12-6. Murray C. Coming Apart, The State of White America, 1960-2010. Crown Forum, Random House. 2012.

12-7. Strenze T. Intelligence and socioeconomic success: A meta-analytic review of longitudinal research. Intelligence. 35(5):401-426, 2007.

12-8. Hegelund ER et al. Low IQ as a predictor of unsuccessful educational and occupational achievement: a register-based study of 1,098,742 men in Denmark 1968-2016. Intelligence. 71:46-53, 2018.

12-9. Bjälkebring P, Peters E. Money matters (especially if you are good at math): Numeracy, verbal intelligence, education, and income in satisfaction judgments. PLoS One. 16(11):e0259331, 2021. PubMed ID: 34818354. [Open]

12-10. Marks GN. Cognitive ability has powerful, widespread and robust effects on social stratification: Evidence from the 1979 and 1997 US National Longitudinal Surveys of Youth. Intelligence. 94:101686, 2022.

12-11. Keuschnigg M, van de Rigt, A, Bol T. The plateauing of cognitive ability among top earners. European Sociological Reviews. 39(5):820-833, 2023. [Open]

12-12. Deary I. Why do intelligent people live longer? Nature. 456(7219):175-176, 2008. PubMed ID: 19005537.

12-13. Batty GD, Deary IJ, Gottfredson LS. Premorbid (early life) IQ and later mortality risk: systematic review. Annals of Epidemiology. 17(4):278-288, 2007. PubMed ID: 17174570.

12-14. Fries J, Peitschnig J. An intelligent mind in a healthy body? Predicting health by cognitive ability in a large European sample. Intelligence. 93:101666, 2022.

12-15. Sánchez-Izquierdo M et al. Intelligence and life expectancy in late adulthood: A meta-analysis. Intelligence. 98:101738, 2023.

12-16. Confucius. Analects. ~500 BCE.

12-17. Plato. The Republic. ~360 BCE.

12-18. Smith A. The Wealth of Nations. 1776.

12-19. Diamond J. Guns, Germs and Steel: The Fates of Human Societies. Vintage Publishing. 2002.

12-20. Clark G. A Farewell to Alms. Princeton University Press. 2007.

12-21. Acemoglu D, Robinson JA. Why Nations Fail, The Origins of Power, Prosperity and Poverty. Crown Publishing Group. 2012.

12-22. Fukuyama F. Political Order and Political Decay. Farrar, Straus and Giroux. 2014.

12-23. Dalio R. The Changing World Order; Why Nations Succeed and Fail. Avid Reader Press, Simon and Schuster. 2021.

12-24. Hanushek EA, Woessmann L. Knowledge capital, growth, and the East Asian miracle. Science. 351(6271):344-345, 2016. PubMed ID: 26798003.

12-25. Lynn R, Fuerst J, Kirkegaard EOW. Regional differences in intelligence in 22 countries and their economic, social and demographic correlates: A review. Intelligence. 69:24-36, 2018.

12-26. Rindermann H, Becker D. Flynn-effect and economic growth: Do national increases in intelligence lead to increases in GDP? Intelligence. 69:87-93, 2018.

12-27. Wahlsten D. Genetic experiments with animal learning: a critical review. Behavioral Biology. 7(2):143-182, 1972. PubMed ID: 4557505.

12-28. Croston R et al. Heritability and evolution of cognitive traits. Behavioral Ecology. 26(6):1447-1459, 2015.

12-29. Heron WT. The inheritance of maze learning ability in rats. Journal of Comparative Psychology. 19(1):77-89, 1935.

12-30. Tryon RC. Genetic differences in maze-learning ability in rats. Yearbook of the National Society for the Study of Education. 39:111-119, 1940.

12-31. Heron WT. The inheritance of brightness and dullness in maze learning ability in the rat. The Journal of Genetic Psychology. 59:41-49, 1941.

12-32. Thompson WR. The inheritance and development of intelligence. Research Publications of the Association for Research in Nervous and Mental Disease. 33:209-231, 1954. PubMed ID: 13246079.

12-33. Bignami G. Selection for high rates and low rates of avoidance conditioning in the rat. Animal Behavior. 13(2):221-227, 1965. PubMed ID: 5835838.

12-34. McClearn GE. Behavioral genetics. Annual Review of Genetics. 4:437-468, 1970. PubMed ID: 4269910.

12-35. Bovet D, Bovet-Nitti F, Oliverio A. Genetic aspects of learning and memory in mice. Science. 163(3863):139-149, 1969. PubMed ID: 5762759.

12-36. Galton F. Hereditary Genius. Macmillan and Co. London. 1869.

12-37. Clark G. The Son also Rises. Princeton University Press. 2014.

12-38. Clark G. The inheritance of social status: England, 1600 to 2022. Proceedings of the National Academy of Science USA. 120(27):e2300926120, 2023. PubMed ID: 37364122. [Open]

12-39. Lee JJ. The heritability and persistence of social class in England. Proceedings of the National Academy of Science USA. 120(29):e2309250120, 2023. PubMed ID: 37406089. [Open]

12-40. Bouchard TJ Jr, McGue M. Familial studies of intelligence: a review. Science. 212(4498):1055-1059, 1981. PubMed ID: 7195071. [Open]

12-41. Eftedal NH et al. Beyond additive genetic effects: Explaining family resemblance in school performance across millions of pairs of Norwegian relatives. Proceedings of the National Academy of Science USA. 122(25):e2419627122, 2025. PubMed ID: 40540601. [Open]

12-42. McClearn GE et al. Substantial genetic influence on cognitive abilities in twins 80 or more years old. Science. 276(5318):1560-1563, 1997. PubMed ID: 9171059. [Open]

12-43. Polderman TJ et al. Meta-analysis of the heritability of human traits based on fifty years of twin studies. Nature Genetics. 47(7):702-709, 2015. PubMed ID: 25985137.

12-44. Loehlin JC. Recent adoption studies of IQ. Human Genetics. 55(3):297-302, 1980. PubMed ID: 7203461.

12-45. Plomin R. Blueprint, How DNA Makes Us Who We Are. MIT Press. 2018.

12-46. Bishop EG et al. Development genetic analysis of general cognitive ability from 1 to 12 years in a sample of adoptees, biological siblings, and twins. Intelligence. 31:31-49, 2003.

12-47. Willoughby EA et al. Genetic and environmental contributions to IQ in adoptive and biological families with 30-year-old offspring. Intelligence. 88:101579, 2021.

12-48. Bouchard TJ Jr et al. Sources of human psychological differences: the Minnesota Study of Twins Reared Apart. Science. 250(4978):223-228, 1990. PubMed ID: 2218526.

12-49. Pederson NL et al. A quantitative genetic analysis of cognitive abilities during the second half of the life span. Psychological Science. 3(6):346-353, 1992.

12-50. Segal, NL. Born Together – Reared Apart, The Landmark Minnesota Twin Study. Harvard University Press. 2012.

12-51. Maulik PK et al. Prevalence of intellectual disability: a meta-analysis of population-based studies. Research in Developmental Disabilities. 32(2):419-436, 2011. PubMed ID: 21236634.

12-52. Zablotsky B et al. Prevalence and Trends of Developmental Disabilities among Children in the United States: 2009-2017. Pediatrics. 144(4):e20190811, 2019. PubMed ID: 31558576. [Open]

12-53. Patel DR et al. A clinical primer on intellectual disability. Translational Pediatrics. 9(Suppl 1):S23-S35, 2020. PubMed ID: 32206581. [Open]

12-54. Online Mendelian Inheritance in Man. (https://www.omim.org/). 2025.

12-55. Ropers HH. Genetics of early onset cognitive impairment. Annual Review of Genomics and Human Genetics. 11:161-187, 2010. PubMed ID: 20822471.

12-56. Musante L, Ropers HH. Genetics of recessive cognitive disorders. Trends in Genetics. 30(1):32-39, 2014. PubMed ID: 24176302.

12-57. Vissers LE, Gilissen C, Veltman JA. Genetic studies in intellectual disability and related disorders. Nature Reviews Genetics. 17(1):9-18, 2016. PubMed ID: 26503795.

12-58. Jansen S, Vissers LELM, de Vries BBA. The Genetics of Intellectual Disability. Brain Sciences. 13(2):231, 2023. PubMed ID: 36831774. [Open]

12-59. Neri G et al. X-linked intellectual disability update 2017. American Journal of Medical Genetics A. 176(6):1375-1388, 2018. PubMed ID: 29696803. [Open]

12-60. Schwartz CE et al. X-Linked intellectual disability update 2022. American Journal of Medical Genetics A. 191(1):144-159, 2023. PubMed ID: 36300573.

12-61. Najmabadi H et al. Deep sequencing reveals 50 novel genes for recessive cognitive disorders. Nature. 478(7367):57-63, 2011. PubMed ID: 21937992.

12-62. Martin HC et al. Quantifying the contribution of recessive coding variation to developmental disorders. Science. 362(6419):1161-1164, 2018. PubMed ID: 30409806. [Open]

12-63. Al-Kasbi G et al. The diagnostic yield, candidate genes, and pitfalls for a genetic study of intellectual disability in 118 middle eastern families. Scientific Reports. 12(1):18862, 2022. PubMed ID: 36344539. [Open]

12-64. Deciphering Developmental Disorders Study. Prevalence and architecture of de novo mutations in developmental disorders. Nature. 542(7642):433-438, 2017. PubMed ID: 28135719. [Open]

12-65. Kaplanis J et al. Evidence for 28 genetic disorders discovered by combining healthcare and research data. Nature. 586(7831):757-762, 2020. PubMed ID: 33057194. [Open]

12-66. Brea-Fernández AJ et al. Trio-based exome sequencing reveals a high rate of the de novo variants in intellectual disability. European Journal of Human Genetics. 30(8):938-945, 2022. PubMed ID: 35322241. [Open]

12-67. Reichenberg A et al. Discontinuity in the genetic and environmental causes of the intellectual disability spectrum. Proceedings of the National Academy of Science USA. 113(4):1098-1103, 2016. PubMed ID: 26711998. [Open]

12-68. Davies G et al. Genome-wide association study of cognitive functions and educational attainment in UK Biobank (N=112 151). Molecular Psychiatry. 21(6):758-767, 2016. PubMed ID: 27046643. [Open]

12-69. Sniekers S et al. Genome-wide association meta-analysis of 78,308 individuals identifies new loci and genes influencing human intelligence. Nature Genetics. 49(7):1107-1112, 2017. PubMed ID: 28530673. [Open]

12-70. Lee JJ et al. Gene discovery and polygenic prediction from a genome-wide association study of educational attainment in 1.1 million individuals. Nature Genetics. 50(8):1112-1121, 2018. PubMed ID: 30038396. [Open]

12-71. Savage JE et al. Genome-wide association meta-analysis in 269,867 individuals identifies new genetic and functional links to intelligence. Nature Genetics. 50(7):912-919, 2018. PubMed ID: 29942086. [Open]

12-72. Okbay A et al. Polygenic prediction of educational attainment within and between families from genome-wide association analyses in 3 million individuals. Nature Genetics. 54(4):437-449, 2022. PubMed ID: 35361970. [Open]

12-73. Kingdom R et al. Rare genetic variants in genes and loci linked to dominant monogenic developmental disorders cause milder related phenotypes in the general population. American Journal of Human Genetics. 109(7):1308-1316, 2022. PubMed ID: 35700724. [Open]

12-74. Chen CY et al. The impact of rare protein coding genetic variation on adult cognitive function. Nature Genetics. 55(6):927-938, 2023. PubMed ID: 37231097. [Open]

12-75. Urpa L et al. Evidence for the additivity of rare and common variant burden throughout the spectrum of intellectual disability. European Journal of Human Genetics. 32(5):576-583, 2024. PubMed ID: 38467730. [Open]

12-76. Kingdom R et al. Genetic modifiers of rare variants in monogenic developmental disorder loci. Nature Genetics. 56(5):861-868, 2024. PubMed ID: 38637616. [Open]

12-77. Demontis D et al. Rare genetic variants confer a high risk of ADHD and implicate neuronal biology. Nature. 649(8098):909-917, 2026. PubMed ID: 41224997. [Open]

12-78. Chung C et al. Comprehensive multi-omic profiling of somatic mutations in malformations of cortical development. Nature Genetics. 55(2):209-220, 2023. PubMed ID: 36635388. [Open]

12-79. Francis DI et al. Comparing saliva and blood for the detection of mosaic genomic abnormalities that cause syndromic intellectual disability. European Journal of Human Genetics. 31(5):521-525, 2023. PubMed ID: 36446895. [Open]

12-80. Denny L, Coles S, Blitz R. Fetal Alcohol Syndrome and Fetal Alcohol Spectrum Disorders. American Family Physician. 96(8):515-522, 2017. PubMed ID: 29094891. [Open]

12-81. Vorgias D, Bynum FD, Bernstein B. Fetal Alcohol Syndrome. In: StatPearls [Internet]. Treasure Island (FL): StatPearls Publishing. 2023. PubMed ID: 28846263. [Open]

12-82. World Health Organization. Preterm Birth. (https://www.who.int/news-room/fact-sheets/detail/preterm-birth). 2023.

12-83. Kroll J et al. Systematic assessment of perinatal and socio-demographic factors associated with IQ from childhood to adult life following very preterm birth. Intelligence. 77:101401, 2019.

12-84. Inder TE, Volpe JJ, Anderson PJ. Defining the Neurologic Consequences of Preterm Birth. New England Journal of Medicine. 389(5):441-453, 2023. PubMed ID: 37530825.

12-85. Mertens A et al. Child wasting and concurrent stunting in low- and middle-income countries. Nature. 621(7979):558-567, 2023. PubMed ID: 37704720. [Open]

12-86. Yang X The global burden, trends, and inequalities of individuals with developmental and intellectual disabilities attributable to iodine deficiency from 1990 to 2019 and its prediction up to 2030. Frontiers in Nutrition. 11:1366525, 2024. PubMed ID: 38953045. [Open]

12-87. Rees J, Castellano S, Andrés AM. Global impact of micronutrients in modern human evolution. American Journal of Human Genetics. 112(10):2538-2561, 2025. PubMed ID: 40934922. [Open]

12-88. McFarland MJ, Hauer ME, Reuben A. Half of US population exposed to adverse lead levels in early childhood. Proceedings of the National Academy of Science USA. 119(11):e2118631119, 2022. PubMed ID: 35254913. [Open]

12-89. Jia Z et al. Prenatal Lead Exposure, Genetic Factors, and Cognitive Developmental Delay. JAMA Network Open. 6(10):e2339108, 2023. PubMed ID: 37870833. [Open]

12-90. Cooper RM, Zubek JP. Effects of enriched and restricted early environments on the learning ability of bright and dull rats. Canadian Journal of Psychology. 12(3):159-164, 1958. PubMed ID: 13573245.

12-91. Nelson CA 3rd et al. Cognitive recovery in socially deprived young children: the Bucharest Early Intervention Project. Science. 318(5858):1937-1940, 2007. PubMed ID: 18096809. [Open]

12-92. Humphreys KL et al. Foster care leads to sustained cognitive gains following severe early deprivation. Proceedings of the National Academy of Science USA. 119(38):e2119318119, 2022. PubMed ID: 36095188. [Open]

12-93. Capron C, Duyme M. Assessment of effects of socio-economic status on IQ in a full cross-fostering study. Nature. 340:552-554, 1989.

12-94. van Ijzendoorn MH, Juffer F, Poelhuis CW. Adoption and cognitive development: a meta-analytic comparison of adopted and nonadopted children's IQ and school performance. Psychological Bulletin. 131(2):301-316, 2005. PubMed ID: 15740423.

12-95. Nisbett RE et al. Intelligence: new findings and theoretical developments. American Psychologist. 67(2):130-159, 2012. PubMed ID: 22233090.

12-96. Kendler KS et al. Family environment and the malleability of cognitive ability: a Swedish national home-reared and adopted-away cosibling control study. Proceedings of the National Academy of Science USA. 112(15):4612-4617, 2015. PubMed ID: 25831538. [Open]

12-97. Koppen-Thulesius LK, Teichmann H. Accelerative trends in intellectual development. British Journal of Social and Clinical Psychology. 11(3):284-294, 1972.

12-98. Lynn R. IQ in Japan and the United States shows a growing disparity. Nature. 297:222-223, 1982.

12-99. Flynn JR. Massive IQ gains in 14 nations: what IQ tests really measure. Psychological Bulletin. 101(2):171-191, 1987.

12-100. Flynn JR. Are We Getting Smarter? Rising IQ in the Twenty-First Century. Cambridge University Press. 2012.

12-101. Bratsberg B, Rogeberg O. Flynn effect and its reversal are both environmentally caused. Proceedings of the National Academy of Science USA. 115(26):6674-6678, 2018. PubMed ID: 29891660. [Open]

12-102. Hegelund ER, Okholm GT, Teasdale TW. The secular trend of intelligence test scores in the present century: the Danish experience. Intelligence. 85:101525, 2021.

12-103. Dworak EM, Revelle W, Condon DM. Looking for Flynn effects in a recent online U.S. adult sample: examining shifts within the SAPA Project. Intelligence. 98:101734, 2023.

12-104. Meisenberg G, Lynn R. Ongoing trends of human intelligence. Intelligence. 96:101708, 2023.

12-105. Flynn JR, Shayer M. IQ decline and Piaget: does the rot start at the top? Intelligence. 66:112-121, 2018.

12-106. Devlin B, Daniels M, Roeder K. The heritability of IQ. Nature. 388(6641):468-471, 1997. PubMed ID: 9242404.

12-107. Bouchard TJ Jr. Genes, evolution and intelligence. Behavior Genetics. 44(6):549-577, 2014. PubMed ID: 24604063.

12-108. Plomin R, von Stumm S. The new genetics of intelligence. Nature Reviews Genetics. 19(3):148-159, 2018. PubMed ID: 29335645. [Open]

12-109. Crabtree GR. Our fragile intellect. Part I. Trends in Genetics. 29(1):1-3, 2013. PubMed ID: 23153596.

Chapter 13 Evolution of Human Intelligence

13-1. Wood B, Harrison T. The evolutionary context of the first hominins. Nature. 470(7334):347-352, 2011. PubMed ID: 21331035.

13-2. White TD et al. Neither chimpanzee nor human, Ardipithecus reveals the surprising ancestry of both. Proceedings of the National Academy of Science USA. 112(16):4877-4884, 2015. PubMed ID: 25901308. [Open]

13-3. Fuss J et al. Potential hominin affinities of Graecopithecus from the Late Miocene of Europe. PLoS One. 12(5):e0177127, 2017. PubMed ID: 28531170. [Open]

13-4. Pugh KD. Phylogenetic analysis of Middle-Late Miocene apes. Journal of Human Evolution. 165:103140, 2022. PubMed ID: 35272113.

13-5. Kun E et al. The genetic architecture and evolution of the human skeletal form. Science. 381(6655):eadf8009, 2023. PubMed ID: 37471560. [Open]

13-6. Ponce de León MS et al. The primitive brain of early *Homo*. Science. 372(6538):165-171, 2021. PubMed ID: 33833119. [Open]

13-7. Lindhout FW et al. A molecular and cellular perspective on human brain evolution and tempo. Nature. 630(8017):596-608, 2024. PubMed ID: 38898293.

13-8. Lancaster MA. Unraveling mechanisms of human brain evolution. Cell. 187(21):5838-5857, 2024. PubMed ID: 39423803. [Open]

13-9. Roberts A, editor. Evolution, the Human Story. Penguin Random House. Second Edition. 2018.

13-10. Wikipedia contributors. List of human evolution fossils. Wikipedia, The Free Encyclopedia. (https://en.wikipedia.org/w/index.php?title=List_of_human_evolution_fossils&oldid=1281844603) (March 26, 2025).

13-11. Paige J, Perreault C. 3.3 million years of stone tool complexity suggests that cumulative culture began during the Middle Pleistocene. Proceedings of the National Academy of Science USA. 121(26):e2319175121, 2024. PubMed ID: 38885385. [Open]

13-12. Ma DD et al. Earliest Prepared core technology in Eurasia from Nihewan (China): Implications for early human abilities and dispersals in East Asia. Proceedings of the National Academy of Science USA. 121(11):e2313123121, 2024. PubMed ID: 38437546. [Open]

13-13. MacDonald K, Scherjon F, van Veen E, Vaesen K, Roebroeks W. Middle Pleistocene fire use: The first signal of widespread cultural diffusion in human evolution. Proceedings of the National Academy of Science USA. 118(31):e2101108118, 2021. PubMed ID: 34301807. [Open]

13-14. Davis R et al. Earliest evidence of making fire. Nature. 649(8097):631-637, 2026. PubMed ID: 41372405.

13-15. Conard NJ, Malina M, Münzel SC. New flutes document the earliest musical tradition in southwestern Germany. Nature. 460(7256):737-740, 2009. PubMed ID: 19553935.

13-16. Meyer M et al. Nuclear DNA sequences from the Middle Pleistocene Sima de los Huesos hominins. Nature. 531(7595):504-507, 2016. PubMed ID: 26976447.

13-17. Dalén L et al. Deep-time paleogenomics and the limits of DNA survival. Science. 382(6666):48-53, 2023. PubMed ID: 37797036. [Open]

13-18. Hess E. Nim Chimpsky, The Chimp Who Would be Human. A Bantam Book. 2008.

13-19. Wikipedia contributors. Koko. Wikipedia, The Free Encyclopedia. (https://en.wikipedia.org/w/index.php?title=Koko&oldid=1269586606) (March 26, 2025).

13-20. Caglayan E et al. Molecular features driving cellular complexity of human brain evolution. Nature. 620(7972):145-153, 2023. PubMed ID: 37468639. [Open]

13-21. Joshy D, Santpere G, Yi SV. Accelerated cell-type-specific regulatory evolution of the human brain. Proceedings of the National Academy of Science USA. 121(52):e2411918121, 2024. PubMed ID: 39680759. [Open]

13-22. Zeberg H, Jakobsson M, Pääbo S. The genetic changes that shaped Neandertals, Denisovans, and modern humans. Cell. 187(5):1047-1058, 2024. PubMed ID: 38367615. [Open]

13-23. Ju XC et al. The activity and expression of adenylosuccinate lyase were reduced during modern human evolution, affecting brain and behavior. Proceedings of the National Academy of Science USA. 2025 Aug 12;122(32):e2508540122, 2025. PubMed ID: 40758872. [Open]

13-24. Darwin C. The Descent of Man. Encyclopedia Britannica Inc. 1871.

13-25. Mangan RJ et al. Adaptive sequence divergence forged new neurodevelopmental enhancers in humans. Cell. 185(24):4587-4603, 2022. PubMed ID: 36423581. [Open]

13-26. Huguet G et al. Effects of gene dosage on cognitive ability: A function-based association study across brain and non-brain processes. Cell Genomics. 4(12):100721, 2024. PubMed ID: 39667348. [Open]

13-27. Karczewski KJ et al. The mutational constraint spectrum quantified from variation in 141,456 humans. Nature. 581(7809):434-443, 2020. PubMed ID: 32461654. [Open]

13-28. Sun KY et al. A deep catalogue of protein-coding variation in 983,578 individuals. Nature. 631(8021):583-592, 2024. PubMed ID: 38768635. [Open]

13-29. Christmas MJ et al. Evolutionary constraint and innovation across hundreds of placental mammals. Science. 380(6643):eabn3943, 2023. PubMed ID: 37104599. [Open]

13-30. Kuderna LFK et al. Identification of constrained sequence elements across 239 primate genomes. Nature. 625(7996):735-742, 2024. PubMed ID: 38030727. [Open]

13-31. Lindblad-Toh K et al. A high-resolution map of human evolutionary constraint using 29 mammals. Nature. 478(7370):476-482, 2011. PubMed ID: 21993624. [Open]

13-32. Halligan DL, Keightley PD. Spontaneous mutation accumulation studies in evolutionary genetics. Annual Review of Ecology, Evolution, and Systematics. 40:151-172, 2009.

13-33. Uchimura A et al. Germline mutation rates and the long-term phenotypic effects of mutation accumulation in wild-type laboratory mice and mutator mice. Genome Research. 25(8):1125-1134, 2015. PubMed ID: 26129709. [Open]

13-34. Katju V, Bergthorsson U. Old Trade, New Tricks: Insights into the Spontaneous Mutation Process from the Partnering of Classical Mutation Accumulation Experiments with High-Throughput Genomic Approaches. Genome Biology and Evolution. 11(1):136-165, 2019. PubMed ID: 30476040. [Open]

13-35. Bao K, Melde RH, Sharp NP. Are mutations usually deleterious? A perspective on the fitness effects of mutation accumulation. Evolutionary Ecology. 36(5):753-766, 2022. PubMed ID: 36245676. [Open]

13-36. Scheinfeldt LB, Tishkoff SA. Recent human adaptation: genomic approaches, interpretation and insights. Nature Reviews Genetics. 14(10):692-702, 2013. PubMed ID: 24052086. [Open]

13-37. Fan S et al. Going global by adapting local: A review of recent human adaptation. Science. 354(6308):54-59, 2016. PubMed ID: 27846491. [Open]

13-38. Johnson KE, Voight BF. Patterns of shared signatures of recent positive selection across human populations. Nature Ecology and Evolution. 2(4):713-720, 2018. PubMed ID: 29459708. [Open]

13-39. Ye S et al. Higher oxygen content and transport characterize high-altitude ethnic Tibetan women with the highest lifetime reproductive success. Proceedings of the National Academy of Science USA. 121(45):e2403309121, 2024. PubMed ID: 39432765. [Open]

13-40. Yasuda N et al. The evolution of surnames: an analysis of their distribution and extinction. Theoretical Population Biology. 5(1):123-142, 1974. PubMed ID: 4818404.

13-41. Skopek J, Leopold T. Who becomes a grandparent – and when? Educational differences in the chances and timing of grandparenthood. Demographic Research. 37:917-928, 2017. [Open]

13-42. Labuda D et al. The effective family size of immigrant founders predicts their long-term demographic outcome: From Québec settlers to their 20th-century descendants. PLoS One. 17(5):e0266079, 2022. PubMed ID: 35507549. [Open]

13-43. Clark G, Cummins N, Curtis M. How did the European Marriage Pattern persist? Social versus familial inheritance: England and Quebec, 1650-1850. Economics and Human Biology. 54:101383, 2024. PubMed ID: 38815501. [Open]

13-44. Kolk M, Skirbekk V. Fading family lines- women and men without children, grandchildren and great-grandchildren in 19th, 20th and 21st Century Northern Sweden. Advances in Life Course Research. 53:100481, 2022. PubMed ID: 36652207. [Open]

13-45. Helgason A et al. A populationwide coalescent analysis of Icelandic matrilineal and patrilineal genealogies: evidence for a faster evolutionary rate of mtDNA lineages than Y chromosomes. American Journal of Human Genetics. 72(6):1370-1388, 2003. PubMed ID: 12721957. [Open]

13-46. Courtiol A et al. Natural and sexual selection in a monogamous historical human population. Proceedings of the National Academy of Science USA. 109(21):8044-8049, 2012. PubMed ID: 22547810. [Open]

13-47. Wiuf C, Hein J. On the number of ancestors to a DNA sequence. Genetics. 147(3):1459-1468, 1997. PubMed ID: 9383085. [Open]

13-48. O'Brien E et al. Founder effect: assessment of variation in genetic contributions among founders. Human Biology. 66(2):185-204, 1994. PubMed ID: 8194843.

13-49. Kaplanis J et al. Quantitative analysis of population-scale family trees with millions of relatives. Science. 360(6385):171-175, 2018. PubMed ID: 29496957. [Open]

13-50. Morral N et al. The origin of the major cystic fibrosis mutation (delta F508) in European populations. Nature Genetics. 7(2):169-175, 1994. PubMed ID: 7920636.

13-51. Motulsky AG. Jewish diseases and origins. Nature Genetics. 9(2):99-101, 1995. PubMed ID: 7719352.

13-52. Waldman S et al. Genome-wide data from medieval German Jews show that the Ashkenazi founder event pre-dated the 14th century. Cell. 185(25):4703-4716, 2022. PubMed ID: 36455558. [Open]

13-53. Wang T et al. Founder effect of a prevalent phenylketonuria mutation in the Oriental population. Proceedings of the National Academy of Science USA. 88(6):2146-2150, 1991. PubMed ID: 2006152. [Open]

13-54. Austerlitz F, Heyer E. Social transmission of reproductive behavior increases frequency of inherited disorders in a young-expanding population. Proceedings of the National Academy of Science USA. 95(25):15140-15144, 1998. PubMed ID: 9844029. [Open]

13-55. Engert JC et al. ARSACS, a spastic ataxia common in northeastern Québec, is caused by mutations in a new gene encoding an 11.5-kb ORF. Nature Genetics. 24(2):120-125, 2000. PubMed ID: 10655055.

13-56. Hästbacka J et al. Identification of the Finnish founder mutation for diastrophic dysplasia (DTD). European Journal of Human Genetics. 7(6):664-670, 1999. PubMed ID: 10482955. [Open]

13-57. Williams KB et al. Clinical characterization of familial hypercholesterolemia due to an amish founder mutation in Apolipoprotein B. BMC Cardiovascular Disorders. 22(1):109, 2022. PubMed ID: 35300601. [Open] .

13-58. Moreau C et al. Deep human genealogies reveal a selective advantage to be on an expanding wave front. Science. 334(6059):1148-1150, 2011. PubMed ID: 22052972. [Open]

13-59. Pluzhnikov A et al. Correlation of intergenerational family sizes suggests a genetic component of reproductive fitness. American Journal of Human Genetics. 2007 Jul;81(1):165-169, 2007. PubMed ID: 17564973. [Open]

13-60. Rohde DL, Olson S, Chang JT. Modelling the recent common ancestry of all living humans. Nature. 431(7008):562-566, 2004. PubMed ID: 15457259. [Open]

13-61. Ohno S. The Malthusian parameter of ascents: what prevents the exponential increase of one's ancestors? Proceedings of the National Academy of Science USA. 93(26):15276-15278, 1996. PubMed ID: 8986801. [Open]

13-62. de Waal F. Our Inner Ape. Penguin Group. 2005.

13-63. Diamond J. The Third Chimpanzee, The Evolution and the Future of the Human Animal. Harper Collins. 1992.

13-64. Kano T. The Last Ape, Pygmy Chimpanzee Behavior and Ecology. Stanford University Press. 1986.

13-65. Lippold S et al. Human paternal and maternal demographic histories: insights from high-resolution Y chromosome and mtDNA sequences. Investigative Genetics. 5:13, 2014. PubMed ID: 25254093. [Open]

13-66. Karmin M et al. A recent bottleneck of Y chromosome diversity coincides with a global change in culture. Genome Research. 25(4):459-466, 2015. PubMed ID: 25770088. [Open]

13-67. Fowler C et al. A high-resolution picture of kinship practices in an Early Neolithic tomb. Nature. 601(7894):584-587, 2022. PubMed ID: 34937939. [Open]

13-68. Blöcher J et al. Descent, marriage, and residence practices of a 3,800-year-old pastoral community in Central Eurasia. Proceedings of the National Academy of Science USA. 120(36):e2303574120, 2023. PubMed ID: 37603728. [Open]

13-69. Gnecchi-Ruscone GA et al. Network of large pedigrees reveals social practices of Avar communities. Nature. 629(8011):376-383, 2024. PubMed ID: 38658749. [Open]

13-70. Seersholm FV et al. Repeated plague infections across six generations of Neolithic Farmers. Nature. 632(8023):114-121, 2024. PubMed ID: 38987589. [Open]

13-71. Henrich J. The WIERDest People in the World. Farrar, Straus and Giroux. 2020. (Chapter 8).

13-72. Neel JV. Lessons from a "primitive" people. Science. 170(3960):815-822, 1970. PubMed ID: 5473413.

13-73. Betzig L. Means, variances, and ranges in reproductive success: comparative evidence. Evolution and Human Behavior. 33(4):309-317, 2012.

13-74. Blaffer Hrdy S. Mother Nature, A History of Mothers, Infants, and Natural Selection. Pantheon Books. 1999.

13-75. Wikipedia contributors. Ismail Ibn Sharif. Wikipedia, The Free Encyclopedia. (https://en.wikipedia.org/w/index.php?title=Ismail_Ibn_Sharif&oldid=1272121588) (March 26, 2025).

13-76. Zerjal T et al. The genetic legacy of the Mongols. American Journal of Human Genetics. 72(3):717-721, 2003. PubMed ID: 12592608. [Open]

13-77. Moore LT et al. A Y-chromosome signature of hegemony in Gaelic Ireland. American Journal of Human Genetics. 78(2):334-338, 2006. PubMed ID: 16358217. [Open]

13-78. Neel JV. On Being Headmen. Perspectives in Biology and Medicine. 23(2), part 1: 277-294, 1980.

13-79. von Rueden CR, Jaeggi AV. Men's status and reproductive success in 33 nonindustrial societies: Effects of subsistence, marriage system, and reproductive strategy. Proceedings of the National Academy of Science USA. 113(39):10824-10829, 2016. PubMed ID: 27601650. [Open]

13-80. Neel JV. Physician to the Gene Pool. John Wiley and Sons. 1994. (pp. 186-187).

13-81. Ross CT et al. Reproductive inequality in humans and other mammals. Proceedings of the National Academy of Science USA. 120(22):e2220124120, 2023. PubMed ID: 37216525. [Open]

13-82. Diamond J. The World Until Yesterday, What Can We Learn from Traditional Societies. Viking Press. 2012.

13-83. Scheidel W. Brother-sister marriage in Roman Egypt. Journal of Biosocial Science. 29(3):361-371, 1997. PubMed ID: 9881142.

13-84. Ceballos FC et al. Human inbreeding has decreased in time through the Holocene. Current Biology. 31(17):3925-3934, 2021. PubMed ID: 34216555. [Open]

13-85. Pearson J et al. Mobility and kinship in the world's first village societies. Proceedings of the National Academy of Science USA. 120(4):e2209480119, 2023. PubMed ID: 36649403. [Open]

13-86. Rodríguez-Varela R et al. Five centuries of consanguinity, isolation, health, and conflict in Las Gobas: A Northern Medieval Iberian necropolis. Science Advances. 10(35):eadp8625, 2024. PubMed ID: 39196943. [Open]

13-87. Casals F et al. Whole-exome sequencing reveals a rapid change in the frequency of rare functional variants in a founding population of humans. PLoS Genetics. 9(9):e1003815, 2013. PubMed ID: 24086152. [Open]

13-88. Lim ET et al. Distribution and medical impact of loss-of-function variants in the Finnish founder population. PLoS Genetics. 10(7):e1004494, 2014. PubMed ID: 25078778. [Open]

13-89. Birchler JA et al. Heterosis. Plant Cell. 22(7):2105-2112, 2010. PubMed ID: 20622146. [Open]

13-90. Hochholdinger F, Baldauf JA. Heterosis in plants. Current Biology. 28(18):R1089-R1092, 2018. PubMed ID: 30253145. [Open]

13-91. Bittles AH, Neel JV. The costs of human inbreeding and their implications for variations at the DNA level. Nature Genetics. 8(2):117-121, 1994. PubMed ID: 7842008.

13-92. Verweij KJ et al. The association of genotype-based inbreeding coefficient with a range of physical and psychological human traits. PLoS One. 9(7):e103102, 2014. PubMed ID: 25062308. [Open]

13-93. Joshi PK et al. Directional dominance on stature and cognition in diverse human populations. Nature. 523(7561):459-462, 2015. PubMed ID: 26131930. [Open]

13-94. Howrigan DP et al. Genome-wide autozygosity is associated with lower general cognitive ability. Molecular Psychiatry. 21(6):837-843, 2016. PubMed ID: 26390830. [Open]

13-95. Ceballos FC et al. Runs of homozygosity: windows into population history and trait architecture. Nature Reviews Genetics. 19(4):220-234, 2018. PubMed ID: 29335644.

13-96. Clark DW et al. Associations of autozygosity with a broad range of human phenotypes. Nature Communications. 10(1):4957, 2019. PubMed ID: 31673082. [Open]

13-97. Yengo L et al. Genomic partitioning of inbreeding depression in humans. American Journal of Human Genetics. 108(8):1488-1501, 2021. PubMed ID: 34214457. [Open]

13-98. Malawsky DS et al. Influence of autozygosity on common disease risk across the phenotypic spectrum. Cell. 186(21):4514-4527, 2023. PubMed ID: 37757828. [Open]

13-99. Antón SC. Natural history of Homo erectus. American Journal of Physical Anthropology. Suppl 37:126-170, 2003. PubMed ID: 14666536. [Open]

13-100. Carotenuto F et al. Venturing out safely: The biogeography of Homo erectus dispersal out of Africa. Journal of Human Evolution. 95:1-12, 2016. PubMed ID: 27260171.

13-101. Reich D. Who We Are and How We Got Here, Ancient DNA and the New Science of the Human Past. Pantheon Books. 2018.

13-102. Olalde I et al. The Beaker phenomenon and the genomic transformation of northwest Europe. Nature. 555(7695):190-196, 2018. PubMed ID: 29466337. [Open]

13-103. Fregel R. The origin of the Carpathian Avar elites revealed. Cell Genomics. 2(6):100143, 2022. PubMed ID: 36778139. [Open]

13-104. Gnecchi-Ruscone GA et al. Ancient genomes reveal origin and rapid trans-Eurasian migration of 7th century Avar elites. Cell. 185(8):1402-1413, 2022. PubMed ID: 35366416. [Open]

13-105. Allentoft ME et al. Population genomics of post-glacial western Eurasia. Nature. 625(7994):301-311, 2024. PubMed ID: 38200295. [Open]

13-106. Fortes-Lima CA et al. The genetic legacy of the expansion of Bantu-speaking peoples in Africa. Nature. 625(7995):540-547, 2024. PubMed ID: 38030719. [Open]

13-107. Kawai Y et al. Exploring the genetic diversity of the Japanese population: Insights from a large-scale whole genome sequencing analysis. PLoS Genetics. 19(12):e1010625, 2023. PubMed ID: 38060463. [Open]

13-108. Milner LS. Hardness of Heart Hardness of Life, The Stain of Human Infanticide. Morris Publishing. 1998.

13-109. de Mause L. in The History of Childhood. de Mause L. editor. Jason Aronson. 1974. (pp. 25-39).

13-110. Pinker S. The Better Angels of our Nature. Penguin Random House. 2011. (Chapter 7).

13-111. Vogel F, Motulsky AG. Human Genetics, Problems and Approaches. Springer. Third Edition. 1996. (Chapter 13).

13-112. Horwitz TB et al. Evidence of correlations between human partners based on systematic reviews and meta-analyses of 22 traits and UK Biobank analysis of 133 traits. Nature Human Behaviour. 7(9):1568-1583, 2023. PubMed ID: 37653148. [Open]

13-113. Crow JF. Basic Concepts in Population, Quantitative, and Evolutionary Genetics. W. H. Freeman and Company. 1986. (Chapter 5).

13-114. Clark G. The inheritance of social status: England, 1600 to 2022. Proceedings of the National Academy of Science USA. 120(27):e2300926120, 2023. PubMed ID: 37364122. [Open]

13-115. Diamond J. The Third Chimpanzee, The Evolution and the Future of the Human Animal. Harper Collins. 1992.

13-116. White M. The Great Big Book of Horrible Things. W.W. Norton and Company. 2011.

13-117. Harari YN. Homo Deus. HarperCollins. 2017.

13-118. Ó Gráda C. Famine, A Short History. Princeton University Press. 2009.

13-119. Wade L. An unequal blow. Science. 368(6492):700-703, 2020. PubMed ID: 32409457. [Open]

13-120. Klunk J et al. Evolution of immune genes is associated with the Black Death. Nature. 611(7935):312-319, 2022. PubMed ID: 36261521. [Open]

13-121. Denevan WM, editor. The Native Population of the Americas in 1492. Second Edition. University of Wisconsin Press. 1992.

13-122. Heuveline P. The boundaries of genocide: Quantifying the uncertainty of the death toll during the Pol Pot regime in Cambodia (1975-79). Population Studies. 69(2):201-218, 2015. PubMed ID: 26218856. [Open]

13-123. Malthus TR. An Essay on the Principle of Population, First Edition. Random House: The Modern Library. 1798.

13-124. Tassier T. The Rich Flee and the Poor Take the Bus. Johns Hopkins University Press. 2024.

13-125. Altmejd A et al. Inequality and COVID-19 in Sweden: Relative risks of nine bad life events, by four social gradients, in pandemic vs. prepandemic years. Proceedings of the National Academy of Science USA. 120(46):e2303640120, 2023. PubMed ID: 37943837. [Open]

13-126. Weber JL. Suggestions for Long Term Goals for Humanity. Ten Generations Press. 2026.

Chapter 14 The Problem: Recent Changes in Evolutionary Forces

14-1. Tatsumoto S et al. Direct estimation of de novo mutation rates in a chimpanzee parent-offspring trio by ultra-deep whole genome sequencing. Scientific Reports. 7(1):13561, 2017. PubMed ID: 29093469. [Open]

14-2. Besenbacher S et al. Direct estimation of mutations in great apes reconciles phylogenetic dating. Nature Ecology and Evolution. 3(2):286-292, 2019. PubMed ID: 30664699.

14-3. Wu FL et al. A comparison of humans and baboons suggests germline mutation rates do not track cell divisions. PLoS Biology. 18(8):e3000838, 2020. PubMed ID: 32804933. [Open]

14-4. Bergeron LA et al. Evolution of the germline mutation rate across vertebrates. Nature. 615(7951):285-291, 2023. PubMed ID: 36859541. [Open]

14-5. Schull WJ, Otake M, Neel JV. Genetic effects of the atomic bombs: a reappraisal. Science. 213(4513):1220-1227, 1981. PubMed ID: 7268429.

14-6. Neel JV et al. The children of parents exposed to atomic bombs: estimates of the genetic doubling dose of radiation for humans. American Journal of Human Genetics. 46(6):1053-1072, 1990. PubMed ID: 2339701. [Open]

14-7. Schmidt L et al. Demographic and medical consequences of the postponement of parenthood. Human Reproduction Update. 18(1):29-43, 2012. PubMed ID: 21989171.

14-8. Aitken RJ. What is driving the global decline of human fertility? Need for a multidisciplinary approach to the underlying mechanisms. Frontiers in Reproductive Health. 6:1364352, 2024. PubMed ID: 38726051. [Open]

14-9. Jónsson H et al. Parental influence on human germline de novo mutations in 1,548 trios from Iceland. Nature. 549(7673):519-522, 2017. PubMed ID: 28959963.

14-10. Goldmann JM, Veltman JA, Gilissen C. De Novo Mutations Reflect Development and Aging of the Human Germline. Trends in Genetics. 35(11):828-839, 2019. PubMed ID: 31610893. [Open]

14-11. Pinker S. The Better Angels of our Nature. Penguin Random House. 2011.

14-12. Ó Gráda C. Famine, A Short History. Princeton University Press. 2009.

14-13. Clark G. The inheritance of social status: England, 1600 to 2022. Proceedings of the National Academy of Science USA. 120(27):e2300926120, 2023. PubMed ID: 37364122. [Open]

14-14. Sunde HF et al. Genetic similarity between relatives provides evidence on the presence and history of assortative mating. Nature Communications. 15(1):2641, 2024. PubMed ID: 38531929. [Open]

14-15. Conley D et al. Assortative mating and differential fertility by phenotype and genotype across the 20th century. Proceedings of the National Academy of Science USA. 113(24):6647-6652, 2016. PubMed ID: 27247411. [Open]

14-16. Schwartz CR, Wang Y, Mare RD. Opportunity and change in occupational assortative mating. Social Science Research. 99:102600, 2021. PubMed ID: 34429208. [Open]

14-17. Zeng TC, Henrich J. Cultural evolution may influence heritability by shaping assortative mating. Behavioral and Brain Sciences. 45:e181, 2022. PubMed ID: 36098444.

14-18. U.S. Census. (https://www.census.gov/library/visualizations/interactive/race-and-ethnicity-in-the-united-state-2010-and-2020-census.html). 2020.

14-19. Bryc K et al. The genetic ancestry of African Americans, Latinos, and European Americans across the United States. American Journal of Human Genetics. 96(1):37-53, 2015. PubMed ID: 25529636. [Open]

14-20. Brazilian Census. (https://g1.globo.com/google/amp/economia/censo/noticia/2023/12/22/censo-2022-cor-ou-raca.ghtml). 2022.

14-21. Parra FC et al. Color and genomic ancestry in Brazilians. Proceedings of the National Academy of Science USA. 100(1):177-182, 2003. PubMed ID: 12509516. [Open]

14-22. Kaplanis J et al. Quantitative analysis of population-scale family trees with millions of relatives. Science. 360(6385):171-175, 2018. PubMed ID: 29496957. [Open]

14-23. Seto KC , Parnell S, Elmqvist T. A global outlook on urbanization. In Urbanization, Biodiversity and Ecosystem Services: Challenges and Opportunities. Editors: Elmqvist T, Fragkias M, Goodness J, Guneralp B, Marcotullio PJ, McDonald RI, Parnell S, Schewenius M, Sendstad M, Seto KC, Wilkinson C. 2013. Chapter 1: pp.1-12.

14-24. Our World in Data. (https://ourworldindata.org/urbanization). March 2025.

14-25. Campbell H et al. Human population structure, genome autozygosity and human health. Genome Medicine. 1(9):91, 2009. PubMed ID: 19804611. [Open]

14-26. Nalls MA et al. Measures of autozygosity in decline: globalization, urbanization, and its implications for medical genetics. PLoS Genetics. 5(3):e1000415, 2009. PubMed ID: 19282984. [Open]

14-27. Colbert SMC et al. Declining autozygosity over time: An exploration in over 1 million individuals from three diverse cohorts. American Journal of Human Genetics. 110(6):1008-1014, 2023. PubMed ID: 37178685. [Open]

14-28. Abdellaoui A et al. Educational attainment influences levels of homozygosity through migration and assortative mating. PLoS One. 10(3):e0118935, 2015. PubMed ID: 25734509. [Open]

14-29. Hochholdinger F, Baldauf JA. Heterosis in plants. Current Biology. 28(18):R1089-R1092, 2018. PubMed ID: 30253145. [Open]

14-30. Penrose LS. Evidence of heterosis in man. Proceedings of the Royal Society London B Biological Sciences. 144(915):203-213, 1955. PubMed ID: 13266803.

14-31. Campbell H et al. Effects of genome-wide heterozygosity on a range of biomedically relevant human quantitative traits. Human Molecular Genetics. 16(2):233-241, 2007. PubMed ID: 17220173.

14-32. Kozieł S, Danel DP, Zaręba M. Isolation by distance between spouses and its effect on children's growth in height. American Journal of Biological Anthropology. 146(1):14-19, 2011. PubMed ID: 21732319.

14-33. McQuillan R et al. Evidence of inbreeding depression on human height. PLoS Genetics. 8(7):e1002655, 2012. PubMed ID: 22829771. [Open]

14-34. Zhu C et al. Hybrid marriages and phenotypic heterosis in offspring: evidence from China. Economics and Human Biology. 29:102-114, 2018.

14-35. Mingroni MA. Resolving the IQ paradox: heterosis as a cause of the Flynn effect and other trends. Psychological Review. 114(3):806-829, 2007. PubMed ID: 17638507.

14-36. Weiss KM. On the number of members of the genus Homo who have ever lived, and some evolutionary implications. Human Biology. 56(4):637-649, 1984. PubMed ID: 6442261.

14-37. Cohen JE. How Many People Can the Earth Support? W. W. Norton and Company. 1995.

14-38. Keinan A, Clark AG. Recent explosive human population growth has resulted in an excess of rare genetic variants. Science. 336(6082):740-743, 2012. PubMed ID: 22582263. [Open]

14-39. Worldometer Population Estimates. (https://www.worldometers.info/world-population/world-population-by-year/). 2025.

14-40. Bailey M, Kaufman AJ. Polygamy in the Monogamous World. Greenwood Publishing Group. 2010.

14-41. Kramer S. Polygamy is rare around the world and mostly confined to a few regions. Pew Research Center. (https://www.pewresearch.org/short-reads/2020/12/07/polygamy-is-rare-around-the-world-and-mostly-confined-to-a-few-regions/#:~:text=Only%20about%202%25%20of%20the,that%20share%20is%20under%200.5%25). 2020.

14-42. Kreyenfeld M, Konietzka D editors. Childlessness in Europe: Contexts, Causes, and Consequences. Springer. (https://library.oapen.org/bitstream/handle/20.500.12657/29707/2017_Book_ChildlessnessInEuropeContextsC.pdf?sequence=1#page=65). 2017. (especially Chapter 2)

14-43. Sobotka T. World's highest childlessness levels in East Asia. Institute for Demographic Studies. Publication number 595. (https://www.ined.fr/fichier/s_rubrique/32255/595_ang.en.pdf). December 2021.

14-44. Reeves RV. Men without children. (https://ofboysandmen.substack.com/p/men-without-children). October 2023.

14-45. Jiang Q et al. Rising trend of childlessness in China: analysis of social and regional disparities with 2010 and 2020 census data. BMJ Open. 13(5):e070553, 2023. PubMed ID: 37236662. [Open]

14-46. Pew Research Center Report. The experiences of U.S. adults who don't have children. (https://www.pewresearch.org/social-trends/2024/07/25/the-experiences-of-u-s-adults-who-dont-have-children/). July 2024.

14-47. Zietsch BP et al. Perfect genetic correlation between number of offspring and grandoffspring in an industrialized human population. Proceedings of the National Academy of Science USA. 111(3):1032-1036, 2014. PubMed ID: 24395780. [Open]

14-48. Labuda D et al. The effective family size of immigrant founders predicts their long-term demographic outcome: From Québec settlers to their 20th-century descendants. PLoS One. 17(5):e0266079, 2022. PubMed ID: 35507549. [Open]

14-49. Clark G, Cummins N, Curtis M. How did the European Marriage Pattern persist? Social versus familial inheritance: England and Quebec, 1650-1850. Economics and Human Biology. 54:101383, 2024. PubMed ID: 38815501. [Open]

14-50. Helgason A et al. A populationwide coalescent analysis of Icelandic matrilineal and patrilineal genealogies: evidence for a faster evolutionary rate of mtDNA lineages than Y chromosomes. American Journal of Human Genetics. 72(6):1370-1388, 2003. PubMed ID: 12721957. [Open]

14-51. Pettay JE et al. Heritability and genetic constraints of life-history trait evolution in preindustrial humans. Proceedings of the National Academy of Science USA. 102(8):2838-2843, 2005. PubMed ID: 15701704 [Open]

14-52. Austerlitz F, Heyer E. Social transmission of reproductive behavior increases frequency of inherited disorders in a young-expanding population. Proceedings of the National Academy of Science USA. 95(25):15140-15144, 1998. PubMed ID: 9844029. [Open]

14-53. Moreau C et al. Deep human genealogies reveal a selective advantage to be on an expanding wave front. Science. 334(6059):1148-1150, 2011. PubMed ID: 22052972. [Open]

14-54. Pluzhnikov A et al. Correlation of intergenerational family sizes suggests a genetic component of reproductive fitness. American Journal of Human Genetics. 2007 Jul;81(1):165-169, 2007. PubMed ID: 17564973. [Open]

14-55. Beaujouan E, Solaz A. Is the Family Size of Parents and Children Still Related? Revisiting the Cross-Generational Relationship Over the Last Century. Demography. 56(2):595-619, 2019. PubMed ID: 30868472 [Open]

14-56. Kolk M, Skirbekk V. Fading family lines- women and men without children, grandchildren and great-grandchildren in 19th, 20th and 21st Century Northern Sweden. Advances in Life Course Research. 53:100481, 2022. PubMed ID: 36652207. [Open]

14-57. Milot E et al. Evidence for evolution in response to natural selection in a contemporary human population. Proceedings of the National Academy of Science USA. 108(41):17040-17045, 2011. PubMed ID: 21969551 [Open]

14-58. Morland P. Tomorrow's People, The Future of Humanity in Ten Numbers. Picador. 2022.

14-59. Sciubba JD. 8 Billion and Counting. W.W. Norton and Company. 2022.

14-60. United Nations World Population Prospects. (https://population.un.org/wpp/). 2025.

14-61. Crow JF. Some possibilities for measuring selection intensities in man. Human Biology. 30(1):1-13, 1958. PubMed ID: 2699600.

14-62. Oeppen J, Vaupel JW. Demography. Broken limits to life expectancy. Science. 296(5570):1029-1031, 2002. PubMed ID: 12004104.

14-63. Johnson S. Extra Life, The Astonishing Story of How We Doubled Our Lifespan. Viking. 2023.

14-64. Gurven M, Kaplan H. Longevity among hunter/gatherers: a cross-cultural examination. Population and Development Review. 33(2):321-365, 2007.

14-65. Clark G. A Farewell to Alms. Princeton University Press. 2007. (Tables 5.1 and 5.2, pp. 93-94)

14-66. Dattani S et al. Child and Infant Mortality. Our World in Data. (https://ourworldindata.org/child-mortality). 2023.

14-67. Corbett S et al. The transition to modernity and chronic disease: mismatch and natural selection. Nature Reviews Genetics. 19(7):419-430, 2018. PubMed ID: 29743650.

14-68. Hill K et al. Mortality rates among wild chimpanzees. Journal of Human Evolution. 40(5):437-450, 2001. PubMed ID: 11322804.

14-69. Hassell JM et al. Morbidity and mortality in infant mountain gorillas (Gorilla beringei beringei): A 46-year retrospective review. American Journal of Primatology. 79(10):e22686, 2017. PubMed ID: 28749595.

14-70. Feudtner C. Bittersweet: Diabetes, Insulin, and the Transformation of Illness. University of Carolina, Chapel Hill. 2003.

14-71. WebMD. Type I Diabetes. (https://www.webmd.com/diabetes/type-1-diabetes). 2024.

14-72. Jonasson JM et al. Fertility in women with type 1 diabetes: a population-based cohort study in Sweden. Diabetes Care. 30(9):2271-2276, 2007. PubMed ID: 17563340. [Open]

14-73. Sjöberg L et al. Fertility in people with childhood-onset type 1 diabetes. Diabetologia. 56(1):78-81, 2013. PubMed ID: 23011355. [Open]

14-74. Wiebe JC et al. Fertility is reduced in women and in men with type 1 diabetes: results from the Type 1 Diabetes Genetics Consortium (T1DGC). Diabetologia. 57(12):2501-2504, 2014. PubMed ID: 25216722. [Open]

14-75. Piret J, Boivin G. Pandemics Throughout History. Frontiers in Microbiology. 11:631736, 2021. PubMed ID: 33584597. [Open]

14-76. Tangye SG et al. Human Inborn Errors of Immunity: 2019 Update on the Classification from the International Union of Immunological Societies Expert Committee. Journal of Clinical Immunology. 40(1):24-64, 2020. PubMed ID: 31953710. [Open]

14-77. Hoffman JI, Kaplan S. The incidence of congenital heart disease. Journal of the American College of Cardiology. 39(12):1890-1900, 2002. PubMed ID: 12084585. [Open]

14-78. van der Linde D et al. Birth prevalence of congenital heart disease worldwide: a systematic review and meta-analysis. Journal of the American College of Cardiology. 58(21):2241-2247, 2011. PubMed ID: 22078432. [Open]

14-79. Alankarage D et al. Identification of clinically actionable variants from genome sequencing of families with congenital heart disease. Genetics in Medicine. 21(5):1111-1120, 2019. PubMed ID: 30293987. [Open]

14-80. Diab NS et al. Molecular Genetics and Complex Inheritance of Congenital Heart Disease. Genes (Basel). 12(7):1020, 2021. PubMed ID: 34209044. [Open]

14-81. Morton SU et al. Genomic frontiers in congenital heart disease. Nature Reviews Cardiology. 19(1):26-42, 2022. PubMed ID: 34272501. [Open]

14-82. Sierant MC et al. Genomic analysis of 11,555 probands identifies 60 dominant congenital heart disease genes. Proceedings of the National Academy of Science USA. 122(13):e2420343122, 2025. PubMed ID: 40127276. [Open]

14-83. Kars ME et al. Deciphering the digenic architecture of congenital heart disease using trio exome sequencing data. American Journal of Human Genetics. 112(3):583-598, 2025. PubMed ID: 39983722. [Open]

14-84. Chien SJ et al. Congenital Heart Diseases Impair Female Fertility. Frontiers in Pediatrics. 9:687276, 2021. PubMed ID: 34336741. [Open]

14-85. Udholm LF et al. Congenital Heart Disease and Fertility: A Danish Nationwide Cohort Study Including Both Men and Women. Journal of the American Heart Association. 12(2):e027409, 2023. PubMed ID: 36648105. [Open]

14-86. Taiber S et al. The Genomics of Auditory Function and Disease. Annual Review of Genomics and Human Genetics. 23:275-299, 2022. PubMed ID: 35667089. [Open]

14-87. Petit C, Bonnet C, Safieddine S. Deafness: from genetic architecture to gene therapy. Nature Reviews Genetics. 24(10):665-686, 2023. PubMed ID: 37173518. [Open]

14-88. Bender RE. The Conquest of Deafness: A History of the Long Struggle to Make Possible Normal Living to Those Handicapped by Lack of Normal Hearing. The Press of Western Reserve University. 1960.

14-89. Blanton SH et al. Fitness among individuals with early childhood deafness: Studies in alumni families from Gallaudet University. Annals of Human Genetics. 74(1):27-33, 2010. PubMed ID: 19930248. [Open]

14-90. Liu X et al. Epidemiological and genetic studies of congenital profound deafness in the general population of Sichuan, China. American Journal of Medical Genetics. 53(2):192-195, 1994. PubMed ID: 7856647.

14-91. Tekin M et al. *GJB2* mutations in Mongolia: complex alleles, low frequency, and reduced fitness of the deaf. Annals of Human Genetics. 74(2):155-164, 2010. PubMed ID: 20201936. [Open]

14-92. Skirbekk V, Bonsang É, Engdahl B. Childhood hearing impairment and fertility in Norway. Scientific Reports. 12(1):402, 2022. PubMed ID: 35013422. [Open]

14-93. Nance WE, Kearsey MJ. Relevance of connexin deafness (DFNB1) to human evolution. American Journal of Human Genetics. 74(6):1081-1087, 2004. PubMed ID: 15079193. [Open]

14-94. Braun DC et al. Deaf intermarriage has limited effect on the prevalence of recessive deafness and no effect on underlying allelic frequency. PLoS One. 15(11):e0241609, 2020. PubMed ID: 33147256. [Open]

14-95. Arnos KS et al. A comparative analysis of the genetic epidemiology of deafness in the United States in two sets of pedigrees collected more than a century apart. American Journal of Human Genetics. 83(2):200-207, 2008. PubMed ID: 18656178. [Open]

14-96. You WP, Henneberg M. Type 1 diabetes prevalence increasing globally and regionally: the role of natural selection and life expectancy at birth. BMJ Open Diabetes Research and Care. 4(1):e000161, 2016. PubMed ID: 26977306. [Open]

14-97. You W, Henneberg M. Modern medical services, a double-edged sword manages symptoms, but accumulates genetic background of cardiovascular diseases: A cross populational analysis of 217 countries. Health Science Reports. 7(1):e1828, 2024. PubMed ID: 38260183. [Open]

14-98. Costa T, Scriver CR, Childs B. The effect of Mendelian disease on human health: a measurement. American Journal of Medical Genetics. 21(2):231-242, 1985. PubMed ID: 4014310.

14-99. Power RA et al. Fecundity of patients with schizophrenia, autism, bipolar disorder, depression, anorexia nervosa, or substance abuse vs their unaffected siblings. JAMA Psychiatry. 70(1):22-30, 2013. PubMed ID: 23147713. [Open]

14-100. Gardner EJ et al. Reduced reproductive success is associated with selective constraint on human genes. Nature. 603(7903):858-863, 2022. PubMed ID: 35322230.

14-101. Liu A et al. Evidence from Finland and Sweden on the relationship between early-life diseases and lifetime childlessness in men and women. Nature Human Behaviour. 8(2):276-287, 2024. PubMed ID: 38110509. [Open]

14-102. Bundy H, Stahl D, MacCabe JH. A systematic review and meta-analysis of the fertility of patients with schizophrenia and their unaffected relatives. Acta Psychiatrica Scandinavica. 123(2):98-106, 2011. PubMed ID: 20958271.

14-103. Neel JV. On Being Headmen. Perspectives in Biology and Medicine. 23(2), part 1: 277-294, 1980.

14-104. Vining Jr DR. Social versus reproductive success: the central theoretical problem of human sociobiology. The Behavioral and Brain Sciences. 9(1):167-216, 1986.

14-105. Hopcroft RL. Sex, status, and reproductive success in the contemporary United States. Evolution and Human Behavior. 27(2):104-120, 2006.

14-106. Nettle D, Pollet TV. Natural selection on male wealth in humans. The American Naturalist. 172(5):658-666, 2008. PubMed ID: 18808363.

14-107. von Rueden CR, Jaeggi AV. Men's status and reproductive success in 33 nonindustrial societies: Effects of subsistence, marriage system, and reproductive strategy. Proceedings of the National Academy of Science USA. 113(39):10824-10829, 2016. PubMed ID: 27601650. [Open]

14-108. Skirbekk V. Fertility trends by social status. Demographic Research. 18:145-180, 2008. [Open]

14-109. Clark G. The Son also Rises. Princeton University Press. 2007. (Figures 7.2 and 7.3, pp. 133-134)

14-110. Clark G. A Farewell to Alms. Princeton University Press. 2014. (Figure 4.4, p. 89; Figure 6.2, p.116; Table 14.4, p.283; Figure 14.7, p.293)

14-111. Van Court M, Bean FD. Intelligence and fertility in the United States: 1912-1982. Intelligence. 9(1):23-32, 1985. PubMed ID: 12267266.

14-112. Retherford RD, Sewell WH. How intelligence affects fertility. Intelligence. 13(2):169-185, 1989.

14-113. Dye, Jane Lawler. Fertility of American Women: 2006. Current Population Reports, P20-558. U.S. Census Bureau. (https://www.test.census.gov/content/dam/Census/library/publications/2008/demo/p20-558.pdf). 2008.

14-114. Hopcroft RL. Sex differences in the relationship between status and number of offspring in the contemporary U.S. Evolution and Human Behavior. 36(2):146-151, 2015.

14-115. Hamilton BE. Total fertility rates, by maternal educational attainment and race and Hispanic origin: United States, 2019. National Vital Statistics Reports; vol 70 no 5. Hyattsville, MD: National Center for Health Statistics. (https://doi.org/10.15620/cdc:105234). 2021.

14-116. Zang E, Sariego C, Krishnan A. The interplay of race/ethnicity and education in fertility patterns. Population Studies. 76(3):363-385, 2022. PubMed ID: 36256449. [Open]

14-117. Hudde A, Cohen PN. Completed U.S. fertility by sex, cohort, and education level. Socius, Sociological Research for a Dynamic World. (https://doi.org/10.1177/23780231241261610). October 2024.

14-118. Grindstaff CF, Balakrishnan TR, Dewit DJ. Educational attainment, age at first birth and lifetime fertility: an analysis of Canadian fertility survey data. Canadian Review of Sociology. 28(3):324-339, 1991.

14-119. DeCicca P, Krashinsky H. The effect of education on overall fertility. Journal of Population Economics. 36:471-503, 2023.

14-120. Hoem JM, Neyer G, Andersson G. Educational attainment and ultimate fertility among Swedish women born in 1955-59. Demographic Research. 14:381-404, 2006. [Open]

14-121. Monstad K, Propper C, Salvanes KG. Education and fertility: evidence from a natural experiment. Scandinavian Journal of Economics. 110(4):827-852, 2008.

14-122. Barthold JA, Myrskylä M, Jones OR. Childlessness drives the sex difference in the association between income and reproductive success of modern humans. Evolution and Human Behavior. 33(6):628-638, 2012.

14-123. Berrington A, Stone J, Beaujouan E. Educational differences in timing and quantum of childbearing in Britain: a study of cohorts born 1940-1969. Demographic Research. 33:733-764, 2015. [Open]

14-124. Beaujouan E, Berghammer C. The gap between lifetime fertility intentions and completed fertility in Europe and the United States: a cohort approach. Population Research and Policy Review. 38:507-535, 2019. [Open]

14-125. Jalovaara M et al. Education, Gender, and Cohort Fertility in the Nordic Countries. European Journal of Population. 35(3):563-586, 2018. PubMed ID: 31372105. [Open]

14-126. Nisén J et al. Educational Differences in Cohort Fertility Across Sub-national Regions in Europe. European Journal of Population. 37(1):263-295, 2020. PubMed ID: 33597840. [Open]

14-127. Lazzari E, Mogi R, Canudas-Romo V. Educational composition and parity contribution to completed cohort fertility change in low-fertility settings. Population Studies. 75(2):153-167, 2021. PubMed ID: 33780319.

14-128. Ciganda D, Lorenti A, Dommermuth L. Microfoundations of the weakening educational gradient in fertility. Population Studies. 79(1):103-122, 2025. PubMed ID: 38700204. [Open]

14-129. Ermisch J. The recent decline in period fertility in England and Wales: Differences associated with family background and intergenerational educational mobility. Population Studies. 78(2):325-339, 2024. PubMed ID: 37285867. [Open]

14-130. Retherford RD, Ogawa N, Matsukura R, Ihara H. Trends in fertility by education in Japan, 1966-2000. (https://www.eastwestcenter.org/sites/default/files/filemanager/Research_Program/Retherford_fertility_Japan2004.pdf). 2004.

14-131. Ghaznavi C et al. Salaries, degrees, and babies: Trends in fertility by income and education among Japanese men and women born 1943-1975-Analysis of national surveys. PLoS One. 17(4):e0266835, 2022. PubMed ID: 35476638. [Open]

14-132. Chen H-Y et al. Dysgenic fertility for intelligence and education in Taiwan. Intelligence. 63:29-32, 2017.

14-133. Yoo SH. Educational differentials in cohort fertility during the fertility transition in South Korea. Demographic Research. 30(53):1463-1494, 2014. [Open]

14-134. Monte LM, Knop B. Men's Fertility & Fatherhood: 2014. Current Population Reports, P70-162, U.S. Census Bureau, Washington, DC. (https://www.census.gov/content/dam/Census/library/publications/2019/demo/P70-162.pdf). 2019.

14-135. Martinez GM, Daniels K, Febo-Vazquez I. Fertility of men and women aged 15–44 in the United States: National Survey of Family Growth, 2011–2015. National Health Statistics Reports; number113. National Center for Health Statistics. (https://www.cdc.gov/nchs/data/nhsr/nhsr113.pdf). 2018.

14-136. Nyström S, Bygren LO, Vining DR Jr. Reproduction and level of intelligence. Scandinavian Journal of Social Medicine. 19(3):187-189, 1991. PubMed ID: 1796252.

14-137. Kolk M, Barclay K. Cognitive ability and fertility among Swedish men born 1951-1967: evidence from military conscription registers. Proceedings of the Royal Society B. 286(1902):20190359, 2019. PubMed ID: 31064299. [Open]

14-138. Kolk M, Barclay K. Do income and marriage mediate the relationship between cognitive ability and fertility? Data from Swedish taxation and conscriptions registers for men born 1951-1967, Intelligence. 84:101514, 2021.

14-139. Bratsberg B, Rogeberg O. Stability and change in male fertility patterns by cognitive ability across 32 birth cohorts. Biology Letters. 19(6):20230172, 2023. PubMed ID: 37376853. [Open]

14-140. Rodgers JL et al. Resolving the debate over birth order, family size, and intelligence. American Psychologist. 55(6):599-612, 2000. PubMed ID: 10892201.

14-141. Livingston G, Parker K, Rohal M. Pew Research Center. Childlessness Falls, Family Size Grows Among Highly Educated Women. Washington, D.C. (https://www.pewresearch.org/wp-content/uploads/sites/20/2015/05/2015-05-07_children-ever-born_FINAL.pdf). 2015.

14-142. Guzzo KB, Loo J. Number of Children Ever Born to Women Aged 40-44, 1980-2022. Family Profiles, FP-23-29. Bowling Green, OH: National Center for Family & Marriage Research. (https://doi.org/10.25035/ncfmr/fp-23-29). 2023.

14-143. Wang M, Fuerst J, Ren J. Evidence of dysgenic fertility in China. Intelligence. 57:15-24, 2016.

14-144. Chen S. The Positive Effect of Women's Education on Fertility in Low-Fertility China. European Journal of Population. 38(1):125-161, 2022. PubMed ID: 35370527 [Open]

14-145. Merz E-M, Liefbroer AC. Cross-national differences in the association between educational attainment and completed fertility. Do welfare regimes matter? Vienna Yearbook of Population Research. 15:95-120, 2017.

14-146. Crabtree GR. Our fragile intellect. Part I. Trends in Genetics. 29(1):1-3, 2013. PubMed ID: 23153596.

14-147. Crabtree GR. Our fragile intellect. Part II. Trends in Genetics. 29(1):3-5, 2013. PubMed ID: 23153597.

14-148. Kong A et al. Selection against variants in the genome associated with educational attainment. Proceedings of the National Academy of Science USA. 114(5):E727-E732, 2017. PubMed ID: 28096410. [Open]

14-149. Beauchamp JP. Genetic evidence for natural selection in humans in the contemporary United States. Proceedings of the National Academy of Science USA. 113(28):7774-7779, 2016. PubMed ID: 27402742. [Open]

14-150. Galton F. Inquiries into Human Faculty and its Development. London: J. M. Dent and Sons, Ltd. New York: E. P. Dutton and Co. Second edition. 1907 (first edition 1883).

14-151. Darwin C. The Descent of Man. Encyclopedia Britannica Inc. 1871.

14-152. Wallace AR. Human selection. Popular Science Monthly. 38:93-106, 1890.

14-153. Fisher RA. The Genetical Theory of Natural Selection. Clarendon Press, Oxford. 1930.

14-154. Muller HJ. Our load of mutations. American Journal of Human Genetics. 2(2):111-176, 1950. PubMed ID: 14771033. [Open]

14-155. Crow JF. The high spontaneous mutation rate: is it a health risk? Proceedings of the National Academy of Science USA. 94(16):8380-8386, 1997. PubMed ID: 9237985. [Open]

14-156. Crow JF. The origins, patterns and implications of human spontaneous mutation. Nature Reviews Genetics. 1(1):40-47, 2000. PubMed ID: 11262873.

14-157. Blau WS. Our Genetic Future. 2023.

14-158. Lynch M. Rate, molecular spectrum, and consequences of human mutation. Proceedings of the National Academy of Science USA. 107(3):961-968, 2010. PubMed ID: 20080596. [Open]

14-159. Lynch M. Mutation and Human Exceptionalism: Our Future Genetic Load. Genetics. 202(3):869-875, 2016. PubMed ID: 26953265. [Open]

14-160. Muller HJ. Human evolution by voluntary choice of germ plasm. Science. 134(3480):643-649, 1961. PubMed ID: 13726704.

14-161. Kondrashov A. James Crow (1916-2012). Nature. 481(7382):444, 2012. PubMed ID: 22281585. [Open]

14-162. Gianola D, Rosa GJ, Allison DB. Humble Thanks to a Gentle Giant (an Obituary for James F. Crow). Frontiers in Genetics. 3:93, 2012. PubMed ID: 22675330. [Open]

14-163. Lynn R, Harvey J. The decline of the world's IQ. Intelligence. 36(2):112-120, 2008.

14-164. Meisenberg G. The reproduction of intelligence. Intelligence. 38(2): 220-230, 2010.

Chapter 15 History of Gene Pool Management

15-1. Bennett RL et al. Genetic Counseling and Screening of Consanguineous Couples and Their Offspring: Recommendations of the National Society of Genetic Counselors. Journal of Genetic Counseling. 11(2):97-119, 2002. PubMed ID: 26141656. [Open]

15-2. Woods CG et al. Quantification of homozygosity in consanguineous individuals with autosomal recessive disease. American Journal of Human Genetics. 78(5):889-896, 2006. PubMed ID: 16642444. [Open]

15-3. Fridman H et al. The landscape of autosomal-recessive pathogenic variants in European populations reveals phenotype-specific effects. American Journal of Human Genetics. 108(4):608-619, 2021. PubMed ID: 33740458. [Open]

15-4. Bittles AH et al. Reproductive behavior and health in consanguineous marriages. Science. 252(5007):789-794, 1991. PubMed ID: 2028254.

15-5. Romeo G, Bittles AH. Consanguinity in the contemporary world. Human Heredity. 77(1-4):6-9, 2014. PubMed ID: 25060264.

15-6. Al-Dewik N et al. Clinical exome sequencing in 509 Middle Eastern families with suspected Mendelian diseases: The Qatari experience. American Journal of Medical Genetics Part A. 179(6):927-935, 2019. PubMed ID: 30919572. [Open]

15-7. Monies D et al. The clinical utility of rapid exome sequencing in a consanguineous population. Genome Medicine. 15(1):44, 2023. PubMed ID: 37344829. [Open] 11.

15-8. El Naofal M et al. The genomic landscape of rare disorders in the Middle East. Genome Medicine. 15(1):5, 2023. PubMed ID: 36703223. [Open]

15-9. Chekroun I et al. Genomics of rare diseases in the Greater Middle East. Nature Genetics. 57(3):505-514, 2025. PubMed ID: 39901015.

15-10. Henrich J. The WEIRDest People in the World. Farrar, Straus and Giroux. 2020. (Chapter 5)

15-11. Wikipedia contributors. Cousin marriage law in the United States. Wikipedia, The Free Encyclopedia. (https://en.wikipedia.org/w/index.php?title=Cousin_marriage_law_in_the_United_States&oldid=1278275234). 2025.

15-12. Vogel F, Motulsky AG. Human Genetics, Problems and Approaches. Springer. Third Edition. 1996. (p. 566)

15-13. Jancar J, Johnston SJ. Incest and mental handicap. Journal of Intellectual Disability Research. 34(6):483-490, 1990. PubMed ID: 2077135.

15-14. Bennett RL et al. Inconsistencies in genetic counseling and screening for consanguineous couples and their offspring: the need for practice guidelines. Genetics in Medicine. 1(6):286-292, 1999. PubMed ID: 11258630. [Open]

15-15. Yengo L, Wray NR, Visscher PM. Extreme inbreeding in a European ancestry sample from the contemporary UK population. Nature Communications. 10(1):3719, 2019. PubMed ID: 31481654. [Open]

15-16. Kano T. The Last Ape, Pygmy Chimpanzee Behavior and Ecology. Stanford University Press. 1986.

15-17. de Waal F. Our Inner Ape. Penguin Group. 2005.

15-18. Samuni L, Langergraber KE, Surbeck MH. Characterization of *Pan* social systems reveals in-group/out-group distinction and out-group tolerance in bonobos. Proceedings of the National Academy of Science USA. 119(26):e2201122119, 2022. PubMed ID: 35727986. [Open]

15-19. Galton DJ. Greek theories on eugenics. Journal of Medical Ethics. 24(4):263-267, 1998. PubMed ID: 9752630. [Open]

15-20. Plato. The Republic. Wordsworth Classic. ~360 BCE.

15-21. Aristotle. Politics. University of North Carolina Press. ~330 BCE.

15-22. Maranto G. Quest for Perfection, The Drive to Breed Better Human Beings. iUniverse.com Inc. 1996. (Chapter 3)

15-23. Plutarch. Life of Lycurgus. Penguin Classics. ~100 CE.

15-24. Darwin C. The Descent of Man. Encyclopedia Britannica Inc. 1871

15-25. Die Lange Kerls. (https://www.lange-kerls.de/en/_0). 2025.

15-26. Morris D. A True Regency Gentleman Had Good Breeding. (https://www.moorgatebooks.com/10/a-true-regency-gentleman-had-good-breeding/). 2015.

15-27. Good Breeding in Pride and Prejudice. (https://prideandprejudicethennow.home.blog/1000-adaptations/blog/830-annotations/good-breeding-in-pride-and-prjudice/). 2024.

15-28. Crow JF. Francis Galton: count and measure, measure and count. Genetics. 135(1):1-4, 1993. PubMed ID: 8224811. [Open]

15-29. Galton F. Hereditary Genius. Macmillan and Co. London. 1869.

15-30. Galton F. Inquiries into human faculty and its development. Macmillan and Co. London. 1883.

15-31. Galton F. Hereditary Talent and Character. Macmillans Magazine. 12(68):157-166, 1865.

15-32. Kevles DJ. In the Name of Eugenics. Alfred A. Knopf. 1985.

15-33. Reilly PR. The Surgical Solution, A History of Involuntary Sterilization in the United States. Johns Hopkins University Press. 1991.

15-34. Comfort N. The Science of Human Perfection. Yale University Press. 2012.

15-35. Reilly PR. Eugenics and Involuntary Sterilization: 1907-2015. Annual Review of Genomics and Human Genetics. 16:351-368, 2015. PubMed ID: 26322647. [Open]

15-36. Armstrong C. Thousands of women sterilised in Sweden without consent. British Medical Journal. 315(7108):563, 1997. PubMed ID: 9302943. [Open]

15-37. Lynöe N. Race enhancement through sterilization, Swedish experiences. International Journal of Mental Health. 36(1):17-25, 2007.

15-38. Amy JJ, Rowlands S. Legalised non-consensual sterilisation - eugenics put into practice before 1945, and the aftermath. Part 2: Europe. The European Journal of Contraception and Reproductive Health Care. 23(3):194-200, 2018. PubMed ID: 29671357.

15-39. Tsuchiya T. Eugenic sterilizations in Japan and recent demands for an apology: a report. Ethics and Intellectual Disability. 3(1):1-4, 1997. PubMed ID: 11656892.

15-40. Amy JJ, Rowlands S. Legalised non-consensual sterilisation - eugenics put into practice before 1945, and the aftermath. Part 1: USA, Japan, Canada and Mexico. The European Journal of Contraception and Reproductive Health Care. 23(2):121-129, 2018. PubMed ID: 29624082. [Open]

15-41. Kemp T. Genetic-hygienic experiences in Denmark in recent years. The Eugenics Review. 49(1):11-18, 1957. PubMed ID: 21260729. [Open]

15-42. Caulfield T, Robertson G. Eugenic policies in Alberta. Alberta Law Review. 35:59-80, 1996.

15-43. Haave P. Sterilization under the Swastika, the case of Norway. International Journal of Mental Health. 36(1):45-57, 2007.

15-44. Paul DB, Spencer HG. The hidden science of eugenics. Nature. 374(6520):302-304, 1995. PubMed ID: 7885463.

15-45. Seeman MV. Sterilization of the mentally ill during the years of World War II in Finland. International Journal of Mental Health. 36(1):58-66, 2007.

15-46. Gawin M. Polish psychiatrists and eugenic sterilization during the interwar period. International Journal of Mental Health. 36(1):67-78, 2007.

15-47. Jeanmonod G, Gasser J, Heller G. Historical aspects of psychiatry in French-speaking Switzerland during World War II. International Journal of Mental Health. 36(1):26-44, 2007.

15-48. East EM. Hidden feeblemindedness. Journal of Heredity. 8(5):215-217, 1917.

15-49. Fisher RA. Modern eugenics. Eugenics Review. 18(3):231-236, 1926. PubMed ID: 21259841. [Open]

15-50. Müller-Hill B. Murderous Science (Tödliche Wissenschaft). Cold Spring Harbor Press. 1984 (German original); 1998 (English translation).

15-51. Rotzoll M et al. The first National Socialist extermination crime, the T4 program and its victims. International Journal of Mental Health. 35(3):17-29, 2006.

15-52. Seeman MV. What happened after T4, starvation of psychiatric patients in Nazi Germany. International Journal of Mental Health. 35(4):5-10, 2006.

15-53. Kessler K. Physicians and the Nazi euthanasia program. International Journal of Mental Health. 36(1):4-16, 2007.

15-54. The Straits Times. Did Mr Lee Kuan Yew create a Singapore in his own image? (https://www.straitstimes.com/singapore/did-mr-lee-kuan-yew-create-a-singapore-in-his-own-image). 2015.

15-55. The News Lens. SINGAPORE: Millionaire Ministers and Systemic Inequality. (https://international.thenewslens.com/article/106583). 2018.

15-56. Chan CK. Eugenics on the rise: a report from Singapore. International Journal of Health Services. 15(4):707-712, 1985. PubMed ID: 4077355. [Open]

15-57. Palen JJ. Fertility and eugenics: Singapore's population policies. Population Research and Policy Review. 5(1):3-14, 1986. PubMed ID: 11649224.

15-58. Krishnan P. Audience agency in a curious instance of failed securitization: public resistance to the Singapore government's eugenics program. Security Dialogue. 55(2):179-196, 2024.

15-59. Maranto G. Quest for Perfection, The Drive to Breed Better Human Beings. iUniverse.com Inc. 1996. (Chapter 9)

15-60. Ombelet W, Van Robays J. Artificial insemination history: hurdles and milestones. Facts Views and Vision in ObGyn. 7(2):137-143, 2015. PubMed ID: 26175891. [Open]

15-61. Plotz D. The Genius Factory, The Curious History of the Nobel Prize Sperm Bank. Random House. 2005.

15-62. Grand View Research. Sperm Bank Market Summary. (https://www.grandviewresearch.com/industry-analysis/sperm-bank-market). 2023.

15-63. Arocho R, Lozano EB, Halpern CT. Estimates of donated sperm use in the United States: National Survey of Family Growth 1995-2017. Fertility and Sterility. 112(4):718-723, 2019. PubMed ID: 31371048. [Open]

Chapter 16 Racism, Elitism, and Other Risks

16-01. Rose S. Darwin 200: Should scientists study race and IQ? NO: Science and society do not benefit. Nature. 457(7231):786-788, 2009. PubMed ID: 19212384.

16-02. Ceci S, Williams WM. Darwin 200: Should scientists study race and IQ? YES: The scientific truth must be pursued. Nature. 457(7231):788-789, 2009. PubMed ID: 19212385.

16-03. Wikipedia contributors. Interracial marriage. Wikipedia, The Free Encyclopedia. (https://en.wikipedia.org/wiki/Interracial_marriage). 2024.

16-04. Alba R. The unappreciated centrality of ethnoracially mixed Americans to the nation's demographic future. Proceedings of the National Academy of Science USA. 121(47):e2415070121, 2024. PubMed ID: 39527739. [Open]

16-05. Bryc K et al. The genetic ancestry of African Americans, Latinos, and European Americans across the United States. American Journal of Human Genetics. 96(1):37-53, 2015. PubMed ID: 25529636. [Open]

16-06. Medina-Muñoz SG et al. Demographic modeling of admixed Latin American populations from whole genomes. American Journal of Human Genetics. 110(10):1804-1816, 2023. PubMed ID: 37725976. [Open]

16-07. Sohail M et al. Mexican Biobank advances population and medical genomics of diverse ancestries. Nature. 622(7984):775-783, 2023. PubMed ID: 37821706. [Open]

16-08 All of Us Research Program Genomics Investigators. Genomic data in the All of Us Research Program. Nature. 627(8003):340-346, 2024. PubMed ID: 38374255 [Open]

16-09. Parra EJ et al. Estimating African American admixture proportions by use of population-specific alleles. American Journal of Human Genetics. 63(6):1839-1851, 1998. PubMed ID: 9837836. [Open]

16-10. Parra FC et al. Color and genomic ancestry in Brazilians. Proceedings of the National Academy of Science USA. 100(1):177-182, 2003. PubMed ID: 12509516. [Open]

16-11. Nunes K et al. Admixture's impact on Brazilian population evolution and health. Science. 388(6748):eadl3564, 2025. PubMed ID: 40373151.

16-12. Livingston G, Brown A. Intermarriage in the U.S. 50 Years After Loving v. Virginia. Pew Research Center. (https://www.pewresearch.org/wp-content/uploads/sites/20/2017/05/Intermarriage-May-2017-Full-Report.pdf). 2017.

16-13. Universal Declaration on the Human Genome and Human Rights. (https://www.unesco.org/en/ethics-science-technology/human-genome-and-human-rights). 1997.

Chapter 17 Possible Solutions to the Gene Pool Problem

17-01. Schull WJ, Otake M, Neel JV. Genetic effects of the atomic bombs: a reappraisal. Science. 213(4513):1220-1227, 1981. PubMed ID: 7268429.

17-02. Neel JV et al. The children of parents exposed to atomic bombs: estimates of the genetic doubling dose of radiation for humans. American Journal of Human Genetics. 46(6):1053-1072, 1990. PubMed ID: 2339701. [Open]

17-03. Bergeron LA et al. Evolution of the germline mutation rate across vertebrates. Nature. 615(7951):285-291, 2023. PubMed ID: 36859541. [Open]

17-04. Kaplanis J et al. Genetic and chemotherapeutic influences on germline hypermutation. Nature. 605(7910):503-508, 2022. PubMed ID: 35545669. [Open]

17-05. Harris K. Evidence for recent, population-specific evolution of the human mutation rate. Proceedings of the National Academy of Science USA. 112(11):3439-3444, 2015. PubMed ID: 25733855. [Open]

17-06. GnomAD population database. (https://gnomad.broadinstitute.org/stats). 2025.

17-07. Kong A et al. Selection against variants in the genome associated with educational attainment. Proceedings of the National Academy of Science USA. 114(5):E727-E732, 2017. PubMed ID: 28096410. [Open]

17-08. Mathieson I. Polygenic prediction of human complex traits using ancient DNA. Current Opinion in Genetics and Development. 95:102396, 2025. PubMed ID: 40886582. [Open]

17-09. Stearns SC et al. Measuring selection in contemporary human populations. Nature Reviews Genetics. 11(9):611-622, 2010. PubMed ID: 20680024.

17-10. Bakkensen JB et al. Childbearing, Infertility, and Career Trajectories Among Women in Medicine. JAMA Network Open. 6(7):e2326192, 2023. PubMed ID: 37498595. [Open]

17-11. Almond D, Cheng Y, Machado C. Large motherhood penalties in US administrative microdata. Proceedings of the National Academy of Science USA. 120(29):e2209740120, 2023. PubMed ID: 37428937. [Open]

17-12. United Arab Emirates, Ministry for Health and Prevention. Genetic testing as part of premarital screening for Emiratis (https://mohap.gov.ae/en/w/genetic-testing-as-part-of-premarital-screening-for-emiratis). 2025.

17-13. Abiib S et al. Qatar's genetic counseling landscape: Current insights and future prospects. Genetics in Medicine Open. 2(Suppl 2):101866, 2024. PubMed ID: 39712968. [Open]

17-14. Gietel-Basten S, Rotkirch A, Sobotka T. Changing the perspective on low birth rates: why simplistic solutions won't work. British Medical Journal. 379:e072670, 2022. PubMed ID: 36379520. [Open]

17-15. Jones, RS. Addressing demographic headwinds in Japan: A long-term perspective. OECD Economics Department Working Papers. No. 1792, OECD Publishing, Paris. (https://doi.org/10.1787/96648955-en). 2024.

17-16. Rowe RG, Daley GQ. Induced pluripotent stem cells in disease modelling and drug discovery. Nature Reviews Genetics. 20(7):377-388, 2019. PubMed ID: 30737492. [Open]

17-17. Cerneckis J, Cai H, Shi Y. Induced pluripotent stem cells (iPSCs): molecular mechanisms of induction and applications. Signal Transduction and Targeted Therapy. 9(1):112, 2024. PubMed ID: 38670977. [Open]

17-18. Yamashiro C et al. Generation of human oogonia from induced pluripotent stem cells in vitro. Science. 362(6412):356-360, 2018. PubMed ID: 30237246. [Open]

17-19. Saitou M, Hayashi K. Mammalian in vitro gametogenesis. Science. 374(6563):eaaz6830, 2021. PubMed ID: 34591639. [Open]

17-20. Murase Y et al. In vitro reconstitution of epigenetic reprogramming in the human germ line. Nature. 631(8019):170-178, 2024. PubMed ID: 38768632. [Open]

17-21. Silver LM. Remaking Eden, How Genetic Engineering and Cloning Will Transform the American Family. Avon Books. 1998.

17-22. Olsson PO et al. Insights from one thousand cloned dogs. Science Reports. 12(1):11209, 2022. PubMed ID: 35778582. [Open]

17-23. Liu Z et al. Cloning of Macaque Monkeys by Somatic Cell Nuclear Transfer. Cell. 172(4):881-887, 2018. PubMed ID: 29958110. [Open]

17-24. Wikipedia contributors. Human cloning. Wikipedia, The Free Encyclopedia. (https://en.wikipedia.org/w/index.php?title=Human_cloning&oldid=1275975050). 2025.

17-25. Crow JF. The origins, patterns and implications of human spontaneous mutation. Nature Reviews Genetics. 1(1):40-47, 2000. PubMed ID: 11262873.

17-26. Horn C. Eve, The Disobedient Future of Birth. Profile Books, Wellcome Collection. 2023.

17-27. Kozlov M. Human trials of artificial wombs could start soon. Here's what you need to know. Nature. 621(7979):458-460, 2023. PubMed ID: 37709976. [Open]

17-28. Dunbar CE et al. Gene therapy comes of age. Science. 359(6372):eaan4672, 2018. PubMed ID: 29326244. [Open]

17-29. Schambach A et al. A new age of precision gene therapy. Lancet. 403(10426):568-582, 2024. PubMed ID: 38006899.

17-30. Church G, Regis E. Regenesis. Basic Books. 2012.

17-31. Mason CE. The Next 500 Years, Engineering Life to Reach New Worlds. MIT Press. 2021.

17-32. Goldstein DB. The End of Genetics, Designing Humanity's DNA. Yale University Press. 2021.

17-33. Gibson DG et al. Creation of a bacterial cell controlled by a chemically synthesized genome. Science. 329(5987):52-56, 2010. PubMed ID: 20488990. [Open]

17-34. James JS et al. The design and engineering of synthetic genomes. Nature Reviews Genetics. 26(5):298-319, 2025. PubMed ID: 39506144.

17-35. Schindler D et al. Design, construction, and functional characterization of a tRNA neochromosome in yeast. Cell. 186(24):5237-5253, 2023. PubMed ID: 37944512. [Open]

17-36. Richardson SM et al. Design of a synthetic yeast genome. Science. 355(6329):1040-1044, 2017. PubMed ID: 28280199. [Open]

17-37. Taghon GJ, Strychalski EA. Rise of synthetic yeast: Charting courses to new applications. Cell Genomics. 3(11):100438, 2023. PubMed ID: 38020966. [Open]

17-38. Chen LG et al. A designer synthetic chromosome fragment functions in moss. Nature Plants. 10(2):228-239, 2024. PubMed ID: 38278952.

Chapter 18 Summary of Part II

18-01. Ord T. The Precipice, Existential Risk and the Future of Humanity. Hachette Books. 2020.

18-02. Posner RA. Catastrophe, Risk and Response. Oxford University Press. 2004.

18-03. Bostrom N, Cirkovic MM, Editors. Global Catastrophic Risks. Oxford University Press. 2008.

18-04. Walsh B. End Times, A Brief Guide to the End of the World. Hachette Books. 2019.

INDEX

ABOUT THE AUTHOR

DR. WEBER holds broad experience in the field of human genetics. For roughly the first half of his career, he worked as the head of a large research laboratory. His discovery of short tandem repeat (microsatellite) DNA polymorphisms and construction of some of the first linkage maps of the human genome were important advances in human genetics. He is the author or coauthor of over 200 journal articles and book chapters. During the second half of his career, Dr. Weber served as founder and CEO of one of the largest clinical DNA testing labs in the US. In this role, he oversaw the testing of about 500,000 patients.

More information about Dr. Weber can be found on the Ten Generations Press web site (www.tengenerationspress.org).

www.ingramcontent.com/pod-product-compliance
Lightning Source LLC
LaVergne TN
LVHW052027200926
848481LV00005B/1608